NO MORE
MONKEY BUSINESS

KEVIN SIMINGTON

© Copyright 2019 Kevin Simington

SMART FAITH PRESS

 A catalogue record for this book is available from the National Library of Australia

All rights reserved. No part of this publication may be reproduced, stored in a retrieval system or transmitted in any form by any means electronic, mechanical, photocopying, recording or otherwise, without the prior written permission of the author.

Unless otherwise specified, Scripture quotations are from the New International Version Bible, copyright © 1973. 1978, 1984, 2011, Zondervan, Grand Rapids, Michigan, USA.

Cataloguing in Publication Data:
TITLE: No More Monkey Business
ISBN: 978-0-6484945-1-5
AUTHOR: Kevin Simington
EDITORS: Sandra Simington, John Crooks, James Barclay
COVER DESIGN: Tatiana Villa (viladesign.net)

*To my daughter, Kylie,
an amazing mother, and a woman of faith,
of whom I am very proud.*

# ACKNOWLEDGMENTS

This book would be a much poorer version of itself without the amazing help of my three editors:

- *My wife, Sandy, who read the first draft and made many corrections and suggestions.*

- *James Barclay, who spent many long hours pouring over the manuscript, fact-checking the technical references and making many extremely helpful suggestions.*

- *John Crooks, who scrutinised the third draft, meticulously checking grammar and technical issues, and making further suggestions.*

All three, through many long hours of tedious editing, have helped to make this book a more polished, technically accurate literary work. I cannot thank all of you enough!

# Table of Contents

Preface .................................................................................. ix

Chapter 1: Fact or Fantasy? ........................................................ 1

Chapter 2: Darwin's Imaginative Theory ...................................... 17

Chapter 3: The Problem with Fossils ........................................... 31

Chapter 4: Rocks and the Age of the Earth .................................. 55

Chapter 5: Dirt, Rocks and Water ............................................... 81

Chapter 6: Dinosaurs and Dragons ............................................. 103

Chapter 7: Irreducible Complexity .............................................. 139

Chapter 8: Impossible Genetics .................................................. 169

Chapter 9: The Origin of Life ...................................................... 199

Chapter 10: Meet Your Ancestors ............................................... 217

Chapter 11: A Little Common Sense Please! ................................ 233

Chapter 12: Science and Faith ................................................... 255

Chapter 13: Evidence for a Creator: Cosmology ........................... 279

Chapter 14: Evidence for a Creator: Intelligent Design .................. 297

Chapter 15: Evidence for a Creator: Historical ............................. 315

Chapter 16: The Light From the Stars ......................................... 333

Chapter 17: The Tide is Turning ................................................. 341

Chapter 18: The Atheistic Agenda .............................................. 347

Chapter 19: Evolution and the Bible ........................................... 359

Other Titles By Kevin Simington ................................................ 379

# **PREFACE**

I have spent over 40 years researching the creation / evolution debate, and have witnessed, over that period, the progressive erosion of evolution's evidence base. One by one, the naïve assumptions upon which Charles Darwin based his speculative theory over a century and a half ago have been put to the sword by the findings of modern science. The fossil record has completely failed to demonstrate the gradual evolution of species. Genetics and microbiology have revealed the impossibility of the supposed processes required for evolution to have occurred. Today, the theory of evolution is in deep crisis, and I believe it stands at the very precipice of wholesale abandonment. After more 150 years of research since Charles Darwin's initial speculations, huge flaws and gaping holes are beginning to emerge in his theory.

This book outlines the fascinating and rapidly accumulating scientific evidence that directly contradicts evolutionary theory and which has led to its abandonment by a growing number of prominent scientists. The evidence presented in this book includes the latest findings in the fields of genetics, physics, microbiology, cosmology, paleontology and geology. This book also presents overwhelming scientific evidence that points to a personal, all-powerful, transcendent God as the creator of our extraordinary universe.

I am aware that there are some who believe in both God and evolution. Theistic evolution refers to the belief that God used evolution as the process for creating biological life. I am not of that opinion. The last chapter of this book proposes a theological explanation of the complete incompatibility of the Bible and evolutionary theory. A clear interpretation of the scriptures does not allow us to have a foot in both camps.

Other readers of this book may be creation sceptics or even atheists, who are convinced of the absolute truth of evolution as it has been popularly disseminated. I ask that you read the evidence presented in this book with an open mind, laying aside, as much as humanly possible, your preconceptions.

There may be others who pick up this book out of curiosity, but who do not believe that the creation / evolution debate is of great consequence. It is my view, however, that the theory of evolution has been a significant factor in the decline of belief in God over recent decades. The philosophy of secular humanism that now pervades the developed world is strongly rooted in the alternate story of origins that evolution provides. It is a story of a purely mechanistic universe, without purpose or meaning. It is a story of a universe without either intentional causality or ultimate accountability. In the evolutionary universe, humans are no more than highly evolved animals who arose from nothing, who may live as they wish, and will return to oblivion when they die. In this sense, the story of evolution is the direct antithesis of the story of the Bible. There are profound issues at stake here.

Dr. David Berlinski, Professor of Mathematics and Molecular Biology at Columbia University, stated:

> *"Darwin's theory of evolution is the last of the great nineteenth-century mystery religions. As we speak, it is now following Freudians and Marxism into the nether regions, and I'm quite sure that Freud, Marx and Darwin are commiserating one with the other in the dark dungeon where discarded gods gather."*
> (H.S. Lipson, Physicist Looks at Evolution, Physics Bulletin 31, 1980, p. 138)

It is my hope and prayer that through reading this book, you will become familiar with the important issues surrounding this debate, and that the evidence will lead you to a firm faith in the transcendent, all-powerful God who created you and calls you back into relationship with himself.

# Chapter 1

# Fact or Fantasy?

Evolution is a theory, not a fact. This is an extremely important distinction. In science, ultimately almost all "knowledge" remains within the realm of theory rather than fact. Even something as well-established as the theory of gravity, with its precise formula, cannot be said to have been proven in an ultimate sense, as we may one day find circumstances in either the micro or macro worlds (such as the sub-atomic quantum realm) where gravity behaves very differently. The formula for gravity may one day need to be amended to incorporate further variables. Thus, while we may colloquially describe gravity as a "fact", its precise nature and operation remain theoretical. Ultimately, all scientific theories remain permanently subject to either verification, amendment or disproof, through the examination of ongoing evidence.

Furthermore, not all scientific theories are worthy of the same degree of respect and credibility. While some theories, such as the law of gravity, are substantiated by clear, observable, repeatable, measurable data, other theories are highly speculative and uncorroborated. The current variety of imaginative theories attempting to explain the origin of the universe are examples of the latter kind of theory. Bubble universes[1], mother-baby universes[2], the multiverse[3], oscillating or cyclical universes[4], and even the creation of our universe by aliens[5] are all highly speculative theories that have been recently proposed by scientists to answer the question of ultimate origin. These theories offer contradictory, often mutually exclusive explanations for the origin of the universe. They are based upon little or no hard evidence, arising primarily from the imaginations of the scientists concerned, and certainly do not warrant the same degree of credibility as does the theory of gravity. Thus, the credibility of scientific theories varies enormously, from highly speculative to largely substantiated, depending upon the amount of observable corroborating evidence, as well as the existence of any contrary evidence.

So, where does the theory of evolution fit into this spectrum? Is it overwhelmingly corroborated by observable evidence? Or is it speculative and largely unsubstantiated?

This book aims to demonstrate the highly speculative nature of the theory of evolution. It will document the ever-shrinking evidence base for the theory, including the rapidly accumulating body of scientific evidence that now directly contradicts its fundamental premises. Far from being a proven "fact", evolution is a theory that is coming under increasing scrutiny and criticism from many respected, highly-credentialled scientists.

The theory of evolution, introduced by Charles Darwin in 1859, proposes that all biological life on earth originated from a single-celled organism and, from there, it developed into the complex and varied life-forms that we see today. This progression from the simple to the complex is supposed to have taken place over billions of years by means of genetic variation and natural selection. This refers to a species' ability to make gradual cumulative changes to its basic structure, in response to environmental and other factors, which eventually results in the formation of new physical features. In this way, it is supposed that an amoeba eventually turned into a giraffe, a fish eventually turned into an elephant, and a single-celled bacteria eventually became a professor of astrophysics at Cambridge University. I am not exaggerating! Ultimately, this is what the theory of evolution proposes!

If you have not previously given much thought to the theory of evolution, the chances are that you have made at least one of the following assumptions:

- The theory of evolution is an established, indisputable fact.
- Only "nutters" and religious "kooks" disagree with it.
- It really doesn't matter whether the theory of evolution is true or not.

Over the course of this book, I hope to disavow you of all three of those assumptions.

**The Theory of Evolution is Not Proven**

There is a common misconception that the theory of evolution is a proven fact, substantiated by a vast body of irrefutable evidence. Ask the average person whether they think evolution is true, and they will often express surprise that you feel the need to even ask the question! The concept that all biological life on earth gradually evolved from simple to complex lifeforms over hundreds of millions of years through the chance processes of natural selection and adaptation is now an entrenched tenet of $21^{st}$ century society. It is an assumed "fact". Not only is the theory accepted unquestioningly, but anyone who dares to propose the biblical creation story as an alternate explanation for the origin of complex biological life is considered to be living in denial. He or she is regarded with the same mocking disdain as are those who still believe that the earth is flat. After all, science has "proven" evolution hasn't it?

Despite this popular impression among the general populace, the theory of evolution is not even *close* to being proven at the scientific level. In fact, there is less supporting evidence for the theory now than there ever has been! As this book will demonstrate, scientific discoveries and developments within many of the major branches of science over recent decades have seriously undermined the previously accepted evidence-base for the theory. Some of the evidence that now directly contradicts the theory of evolution, and which will be examined in detail in this book, includes:

- The fossil record, which fails to support the concept of the gradual development of increasingly complex species.

- The vast amounts of complex genetic information (DNA) within almost every cell of every living organism, which could not possibly have been created through random evolutionary processes.

- The now widely recognised genetic impossibility of random evolutionary processes creating the huge amounts of new genetic information (DNA) necessary for the creation of new physical features.

- The now-acknowledged fact that natural selection (small variations within a species - which can be observed today throughout the animal and plant world) does *not* offer a viable means whereby one species can evolve into a completely *different* species with *new* physical characteristics.

- The inability of evolution to explain how the first living cells came into existence from non-living matter.

- The extraordinary complexity of the single cell, comprised of hundreds of complex processes, carried out by hundreds of microscopic, robot-like molecular "machines', which could not possibly have been created through random evolutionary processes.

In response to these and many other recent scientific developments, a growing number of respected scientists are voicing serious concerns regarding the theory, and some are abandoning it altogether. Conscientious objectors to evolution have been growing in number and becoming more vocal for over 50 years:

- In 1967, in response to the growing understanding of genetics, a scientific symposium was held at the Wistar Institute, a biomedical science research centre in Philadelphia, where the plausibility of evolution was challenged because of the now recognised impossibility of new physical features being created by random genetic adaptation.[6]

- In 1973, renowned palaeontologist, Dr. Barbara J. Stahl, drew world attention to the failure of the fossil record to provide any unequivocal evidence for evolution, in her acclaimed book, *"Vertebrate History; Problems in Evolution"*.[7]

- In November 1980, at the Natural History Museum in Chicago, a large number of the world's leading geneticists and other scientists held a seminar to consider the issue of whether the small changes within a species, sometimes referred to as "micro-evolution", can lead to the big changes necessary for Darwinian evolution ("macro-evolution" - one species changing into a *brand-new* species, with new physical features). The findings of the conference were reported in the next issue of "Science" magazine, which stated;

*"The central question of the Chicago conference was whether the mechanisms underlying micro-evolution can be extrapolated to explain the supposed phenomena of macro-evolution. At the risk of doing violence to the opinions of some of the scientists at the meeting, the answer was a clear 'No'."[8]*

In other words, in 1980, a conference of the world's leading geneticists concluded that evolution is not genetically possible!

- In 1980, in response to the findings of the Chicago Conference, Dr. Ken Ham stated;

*"If they had known about genetics in Darwin's day, the theory of evolution would never have gotten off the ground."[9]*

- In 1986, molecular biologist, Dr. Michael Denton, in his landmark book, *"Evolution: A Theory in Crisis"*, outlined the newly emerged understanding of the irreducible complexity of the single cell; that even the simplest cell is a complex bio-factory of hundreds of interdependent molecular components, all of which need to be simultaneously existent and functional for the first cell to be alive. The theory of evolution does not have a viable explanation as to how these complex components could have sprung into existence simultaneously.[10]

- In the 1990s, the huge amounts of information in DNA within cells (which we will examine in a Chapter 7) became a topic of great discussion. How could the 3.2 billion pieces of encoded information within ***every*** DNA strand in almost ***every*** cell of the human body have come into existence through natural processes? Where did this information come from? Evolution simply cannot account for this. Physicist, Dr. Lee Spetner, highlighted these serious problems in his book, *"Not By Chance: Shattering The Modern Theory of Evolution"*, in 1997.[11] This extremely problematic issue has had evolutionists scratching their collective heads ever since. For example, Professor Werner Gitt, of the German Institute of Physics, published a critical book in 2006, entitled, *"In The Beginning Was Information"*.[12] This book outlined the complete inability of random evolutionary processes to produce the vast quantity of complex information within DNA. The existence of DNA is a huge problem for the theory of evolution!

- In 2002, in response to the mounting scientific evidence contradicting the theory of evolution, many of the world's leading scientists began to call for a symposium to determine the ongoing validity of the theory of evolution. As a response, in that same year, an international organisation of scientists was formed, called CESHE (Cercle d'Etudes Scientifique et Historique), headquartered in France. It was a voluntary organisation comprised of many highly respected scientists, whose purpose was to determine whether the theory of evolution can still be considered to be a valid scientific theory. After a period of intense scrutiny and rigorous evaluation of all the evidence, this was their conclusion:

*"The theory of evolution is not supported by science. Many scientists have accepted the theory because they assume it to be an established scientific fact. Those scientists who have investigated it, however, find that evolution is a belief, not a science."*[13]

- In July 2008, as the scientific integrity of the theory of evolution continued to unravel, a conference of the world's leading evolutionary scientists was held in Altenberg, Austria. The purpose of the conference was to discuss the growing realisation that if natural selection (the slight variations **within** a species - which is an observable and undeniable process) cannot produce **new** species with completely **new** physical features, then Darwin's theory is dead. The conference could not come up with a viable explanation of how natural selection could achieve this, given our current knowledge of genetics. After the conference, Dr. Jerry Fodor, of Rutgers University, is quoted as saying, *"Basically I don't think anybody knows how evolution works."*[14]

These few examples represent the tip of the iceberg regarding the growing dissatisfaction with the theory of evolution among the scientific community. Evolution is **far** from proven and an increasing number of respected scientists are daring to voice their disbelief.

In 2001, respected Australian scientist, Dr. John Ashton, published the landmark book, *"In Six Days; Why Fifty Scientists Choose To Believe in Creation"*.[15] His book contains fifty chapters, each written by a different Ph.D. scientist. Each of them provides extensive scientific arguments for their view that the theory of evolution is no longer scientifically tenable, and they explain the growing scientific evidence supporting creation. The contributing scientists are highly regarded internationally, and come from a wide range of fields including biology, chemistry, biochemistry, genetics, physics, zoology, astronomy, meteorology, engineering and botany.

In 2012, Dr John Ashton published the book, *"Evolution Impossible: 12 Reasons Why Evolution Cannot Explain The Origin of Life on Earth"*. The book provides a comprehensive discussion of the profound scientific flaws that have become apparent in the theory of evolution in recent years.

In 2014, respected geneticist, Dr. Casey Luskin, contributed a chapter to the book, *"More Than Myth"*, entitled *"The Top Ten Scientific Problems with Biological and Chemical Evolution"*.[16] The chapter received world-wide attention, prompting many other scientists to voice similar concerns.

In the area of cosmology (the study of the stars and planets), many scientists are concluding that evolutionary theory offers no explanation for the origin of the universe itself. Recent discoveries in cosmology have shown that the universe actually had a beginning; that at some point in the past it simply came into existence. How could the entire physical universe spring into existence from nothing? There is no known natural cause which could account for this. Only something *outside* of nature, something *super*natural, could have caused this. Commenting on the growing number of scientists who now concede that the universe must have a supernatural cause, astrophysicist, Dr Hugh Ross, Director of Observations at Royal Astronomical Society, Vancouver, states;

> *"Astronomers who do not draw theistic or deistic conclusions are becoming rare, and even the few dissenters hint that the tide is against them. Geoffrey Burbidge, of the University of California at San Diego, complains that his fellow astronomers are rushing off to join 'the First Church of Christ of the Big Bang.'"*[17]

In his book, *"God, Science and Evolution"*, Prof. E.H. Andrews wrote;

> *"Speaking as a scientist, I believe that in another 20 years the theory of evolution will have been totally discredited, purely on scientific grounds. The enormous gaps in the theory are beginning to emerge – not, of course, in the popular versions of evolution, but in the findings of scientists who are studying these matters at depth."*[18]

## Scientific Prejudice

Despite the growing discontent with the concept of evolution within the scientific community, there remains a significant number of influential scientists who refuse to relinquish the theory. They cling doggedly to evolutionary theory, refusing to acknowledge the obvious flaws that have emerged in recent years. Furthermore, some evolutionary scientists in positions of influence have been known to unfairly target dissenting scientists. Dr. John F. Ashton, in his book, *"Evolution Impossible"*, cites the recent case of Dr. Gavriel Avital, the chief scientist for the Israeli Education Ministry, who was sacked for voicing his doubts concerning the validity of the theory of evolution.[19] Similarly, the 2008 documentary film, *"Expelled: No Intelligence Allowed"*[20], included interviews with a number of scientists who had been sacked from their positions because they expressed disbelief in evolution. The film presented evidence of the blindly dogmatic allegiance to evolutionary theory within academia and the refusal of those in positions of authority to allow dialogue regarding alternate viewpoints.

The reason for this kind of intransigence (unwillingness to change one's view) is an underlying atheistic philosophy. The theory of evolution does away with the need for a Creator, by providing an explanation for the origin of life based upon purely natural processes. This is an extremely appealing theory for the determined atheist, and scientists who align themselves with this philosophical viewpoint will fight vehemently to defend it.

## Media Bias

The general public's mistaken impression that evolution is irrefutably proven is largely the result of glaring bias in both the education system and popular media. I will discuss the bias within education in Chapter 3, *"The Problem With Fossils"*. In regard to the media, there is significant evidence of evolutionary bias in general reporting and in the production of documentaries and films. Cameron Horn's 1997 book, *"Science V*

*Truth*"[21], provides a disturbing exposé of evolutionary media bias within the Australian media. Citing hundreds of specific examples, Horn reveals a shocking litany of intimidation, censorship and misrepresentation of the creationist viewpoint and the shameless promotion of evolutionary and atheistic philosophy through unbalanced reporting.

A classic case of this was the July 17, 1997 episode of the ABC Science program, *Quantum*. The producers of the program decided to dedicate part of this episode to the creation / evolution debate. Researchers for the program contacted several scientists from the Creation Science Foundation, asking for interviews, with the promise that the resulting segment would be fair and equitable. Here is what resulted:

- When the segment finally aired, it was given the title, "*Telling Lies For God*" and was totally disparaging towards the creationist viewpoint.

- Of the 28 minutes total broadcast time, 20 minutes was dedicated to pro-evolutionary arguments and only 7 minutes was dedicated to the creationist arguments (1 minute was neutral).

- The creationists arguments were heavily redacted, with their most powerful arguments omitted entirely. The portions of their comments that were featured were little more than set ups for the evolutionary arguments.

- The evolutionary scientists were always referred to by their first names, while the creation scientists were referred to by their surnames.

- The video imagery was also shockingly biased. Evolutionists were filmed in attractive outdoor settings - beaches and hillsides - as well as footage of them lecturing to large numbers at university. By comparison, most of the footage of the creationists was shot in church with excruciatingly out-of-tune singing. Cameron Horn, in his book, "Science V Truth", points out further imbalance in the filming: *"The only outdoor shots of creationists resulted from the ABC unit staking out the home of Peter and Cathy Sparrow, whose Creation Bus was then 'buzzed' by the TV crew at high speed on the open highway. There was also an attempt to film a meeting through a window without permission."*[22]

This kind of blatant bias and unbalanced reporting permeates almost every instance of media coverage of the creation / evolution debate, not only in Australia but in the western world generally. The Australian ABC's TV show, "Q and A", is a serial offender, regularly stacking the panel with an array of highly qualified atheists and evolutionists and pitting them against a single "out-gunned" creationist.

A classic case was the 8th March 2010 episode of "Q and A", entitled *"God, Science and Sanity"*, where the producers pitted Senator Steve Fielding, the federal parliamentary leader of the Family First Party, against Dr. Richard Dawkins, a qualified scientist and outspoken atheist and evolutionist. Senator Fielding, a Bible believing Christian, was the sole representative of the creationist viewpoint on the panel, and was constantly asked to defend his position in response to the technical, scientific onslaught from Dr. Dawkins. Senator Fielding repeatedly responded by confessing that he was just an ordinary Christian without scientific training, and that he was not able to comment on the technical aspects of Dr. Dawkins' arguments. On more than one occasion Senator Fielding expressed surprise that he was being asked to comment on scientific matters when he was unqualified to do so, and expressed puzzlement as to why the producers hadn't invited a properly qualified creation scientist to participate on the panel. Why indeed! The answer, it appears, is that the producers had absolutely *no* desire to arrange an even-handed debate, but simply wanted to feed another Christian to the lions!

Media imbalance is also seen in the disproportionate number of evolutionary documentaries that are produced when compared with documentaries that offer a creationist perspective. Subscription channels such as National Geographic and Discovery, as well as free to air TV stations, are replete with documentaries and science programs that portray evolution as an indisputable fact. One can only surmise that there is a strong evolutionary / atheistic bias among those responsible for programming and budget allocation.

Given this extraordinary bias towards evolution in the media, it is no wonder that the general public are convinced of its truth. But for anyone who takes the time and effort to peel back the curtain of popular media programming and actually investigate the hard evidence, a very different picture emerges. While popular media gives the impression of overwhelming consensus within the scientific community regarding evolution, the reality is that there is strong and growing disagreement regarding some of the fundamental assumptions of evolution. While the impression is given of complete confidence, the reality is an increasing atmosphere of doubt and uncertainty. And instead of a vast body of incontestable supporting evidence, the truth is that the evidence-base for evolution is dwindling rapidly, to the point where the theory is in crisis.

As Prof. E.H. Andrews states,

> *"The popular impression is given that evolution is scientifically proven. This view is terribly biased and ignores the yawning chasms in the theory which make it unacceptable to me as a scientist."[23]*

## DOES IT REALLY MATTER?

What's all the fuss about? Does it really matter whether evolution is true or not? I regularly come into contact with people who express the view that the evolution / creation debate is irrelevant and a waste of time.

Their attitude is, *"Let's get on with living life now, rather than engaging in debate about what may or may not have happened in the past! There are much more important issues that need to be addressed in the present!"*

While I understand this viewpoint, I strongly disagree. Evolutionary theory has been one of the driving forces behind the dramatic increase in atheism and agnosticism within Western society. By providing an explanation for the origin and development of life that sidesteps the necessity for a Creator, evolutionary theory has provided people with the perfect excuse to dismiss God and to embrace a secular approach to life in which they no longer see themselves as ultimately accountable to a supreme moral being. This has had a profound impact upon societal morals, including the advent of relativism (the idea that there is no absolute right or wrong) and pluralism (the concept that all views are equally valid). It has also resulted in a dramatic decline in the number of people who profess some form of Christian belief, as indicated by census data and other sociological surveys. (For example, Christian affiliation in Australia has declined from 87% in 1947 to 51% in 2016, and total religious affiliation has declined from 88% to 60% over the same period.[24])

Dr. Richard Dawkins, the renowned evolutionist and outspoken atheist, commented:

> *"Darwin made it possible to be an intellectually fulfilled atheist."*[25]

In response to this downward slide of faith and morals, however, there are many voices now daring to speak up. They point out the glaring faults in the theory of evolution, and the growing body of contrary evidence. They identify the impressive and rapidly accumulating evidence for intelligent design within the universe and the existence of incredibly complex biological features which cannot be explained by natural processes. They dare to speak of the scientific evidence for a Creator.

The evolution / creation debate is not an irrelevance. It is foundational to our sense of meaning and purpose. Because if evolution is *not* true, if we did not evolve via random processes from primordial slime, then we are left with the only logical, viable, alternate explanation; that we were created by an infinite, all-powerful God. And if that is true, it changes everything!

---

**ENDNOTES - Chapter 1**

---

[1] Proposed by Paul Steinhardt and Alexander Vilenkin, in 1983. https://en.wikipedia.org/wiki/Eternal_inflation

[2] Stephen Hawking, "Black Holes and Baby Universes", Random House Publishers, 1994.

[3] Greene, Brian (24 January 2011). "A Physicist Explains Why Parallel Universes May Exist". npr.org(Interview). Interviewed by Terry Gross. Archived from the original on 13 September 2014. Retrieved 12 September 2014.

[4] P. J. Steinhardt, N. Turok (2001). "Cosmic Evolution in a Cyclic Universe". Physical Review D. 65 (12): 126003. arXiv:hep-th/0111098.

[5] Video Clip of interview between Ben Stein and Dr. Richard Dawkins, "Richard Dawkins Believes Extraterrestrials Created Man." https://www.youtube.com/watch?v=AiVoS78lNqM

[6] P.S. Moorhead & M.M. Kaplan, "Mathematical Challenges to the Neo-Darwinian Interpretation of Evolution", The Wistar Institute Symposium Monograph No.5, Philadelphia, P.A., Wistar Institute Press, 1967).

[7] Barbara J. Stahl, "Vertebrate History; Problems in Evolution", New York, McGraw-Hill, 1973.

[8] Roger Lewin, "Science" Journal, Vol. 210(4472), 1980, pp.883-887

[9] Ken Ham, "The Evolution Tapes", 1980

[10] Michael Denton, "Evolution: A Theory in Crisis", Bethesda, Adler & Adler, 1986.

[11] Lee M. Spetner, "Not By Chance: Shattering The Modern Theory of Evolution", New York, Judaica Press, 1997.

[12] Werner Gitt, "In The Beginning Was Information", Green Forest, A.R., Master Books, 2006.

[13] Cercle D'études Scientifique et Historique, https://ceshe.fr/revue-sf-1.html#p=3, also quoted in "Evolution; Fact or Belief?", Creation Science Foundation.

[14] Susan Mazur, The Altenberg 16: "An Expose of the Evolutionary Industry", Berkeley, CA, North Atlantic Books, 2010.

[15] John F. Ashton, "In Six Days: Why Fifty Scientists Choose To Believe in Creation", Green forest, AR, Masterbooks, 2001.

[16] Casey Luskin's chapter, "The Top Ten Scientific Problems with Biological and Chemical Evolution" in the book, "More than Myth" (Chartwell Press, 2014).

[17] Hugh Ross, *"The Creator and The Cosmos"*, Navpress, 2001, pp.108-112

[18] E.H. Andrews, "God, Science and Evolution". Creation Life Publishing, 1981, p.90

[19] John F. Ashton, op. cit., p.22

[20] Directed by Nathan Frankowski, "Expelled: No Intelligence Allowed", 2008.

[21] Cameron Horn, "Science V Truth", Fuzcapp productions, Sydney, 1997, p.276.

[22] Cameron Horn, "Science V Truth", Fuzcapp productions, Sydney, 1997, p.276.

[23] E.H. Andrews, op. cit.

[24] Census data from the Australian Bureau of Statistics, https://www.abs.gov.au/ausstats/abs@.nsf/Lookup/by+Subject/2071.0~2016~Main+Features~Religion+Data+Summary~70

[25] Dawkins R., *The Blind Watchmaker*, Penguin, London, p. 6, 1991

Chapter 2

# Darwin's Imaginative Theory

Most people assume that Charles Darwin's original theory was formed on the basis of substantial scientific evidence, but this is not the case at all. Darwin's theory of the gradual evolution of species through natural selection arose mainly out of imaginative speculation, with very little corroborated, scientific evidence. Darwin admitted as much in a letter to a fellow biologist at the time:

> *"I am quite conscious that my speculations run quite beyond the bounds of true science."*[1]

Darwin openly admitted that his theory was primarily a work of philosophical speculation, rather than a scientific theory. So, what was his theory, and what evidence led him to propose it?

## THE THEORY

Darwin began voicing his speculative ideas in 1838, and formalised them in his now famous book, *"On The Origin of Species"*, in 1859. His theory proposed that all biological life on earth evolved gradually over time, from a single, simple biological ancestor. Thus, all biological species, from worms to humans, are supposedly related to each other and developed along divergent lines through a process he termed "natural selection" (a term which he borrowed from previous scientists such as Edward Blythe, 1810 -1873).[2] Darwin proposed that environmental factors produced slight modifications within successive generations of a species. These modifications supposedly accumulated over time until, eventually, the changes were so significant that an entirely new species emerged. Bacteria grew eyes, gills, fins and internal organs and turned into fish. Fish grew legs and air-breathing lungs and emerged from the sea. The newly emerged fish gradually changed into ants, beetles, mice, dogs and elephants. Reptiles grew wings and became birds. In this way, Darwin imagined a "tree of life" where all living species are related and evolved from a common ancestor. Darwin wrote;

> *"Therefore, I should infer from analogy that probably all organic beings which have ever lived on this earth have descended from one primordial form, into which life was first breathed."*[3]

The following hand-drawn diagram of the tree of life is found in one of Darwin's notebooks:

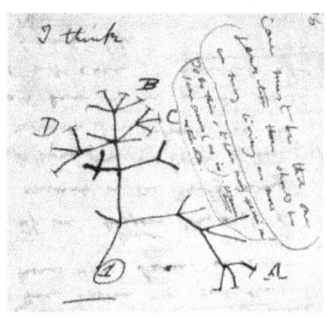

# Darwin's Imaginative Theory

**THE EVIDENCE**

The purely speculative nature of Darwin's tree of life is indicated by Darwin's admission at the tope of the page, "*I think*". In fact, nowhere in any of his notes does Darwin ever give **any** hard evidence for his imaginary tree of life. It is a product of pure imaginative wishful thinking! Today, even many evolutionists are rejecting Darwin's tree of life as overly simplistic. For example, the cover story in the 24th January 2009 edition of the scientific journal, "*New Scientist*", made the startling admission;

> "*The tree of life has turned out to be a figment of our imagination.*"[4]

The imaginative nature of Darwin's speculations is revealed by the language contained in his "*On The Origin of Species*". Throughout the book, instead of appealing to hard evidence, he appeals to his own imagination. Consider the following list of words or phrases and their frequency used by Darwin in the 1872 edition of his book[5]:

- "*may*" (642 times)
- "*might*" (203 times)
- "*probably*" (182 times)
- "*perhaps*" (63 times)
- "*I think*" (43 times)

Here is an example of Darwin's imaginative speculation:

> "*In order to make it clear how, **as I believe**, natural selection acts, I must beg permission to give one or two **imaginary** illustrations. Let us suppose that the fleetest prey, a deer for instance, had from any change in the country increased in numbers, or that*

> *other prey had decreased in numbers. ... Under such circumstances the swiftest and slimmest wolves would have the best chance of surviving, and so be preserved or selected. ... I speak of the slimmest individual wolves, and not of any single strongly-marked variation having been preserved."*[6]

Interestingly, most scientists now readily acknowledge the flawed logic of Darwin's example of wolves, because wolves hunt in packs, not individually, so an individual wolf's survival is not predicated upon whether it is fast or slow, slim or chubby!

Darwin's *"On The Origin of Species"* is full of this kind of evidence-less, flawed logic. In fact, in his introduction to the book, Darwin laments his inability to support his theory with hard evidence, freely admitting that he can produce no paleontological evidence (fossils) of transitional forms between species.[7] He laments this lack of evidence in greater detail in Chapter 6, where he states;

> *"As, by this theory, innumerable transitional forms must have existed, why do we not find them embedded in countless numbers in the earth's crust? ... Why, if species have descended from other species by insensible fine gradations, do we not everywhere see innumerable transitional forms?"*[8]

Darwin remained perplexed about this complete lack of fossil evidence, stating in that same chapter;

> *"I can give no satisfactory answer... Nature may almost be said to have guarded against the discovery of her transitional or linking forms."*[9]

Of course, a much more plausible explanation is that transitional forms simply do not exist!

At one point, Darwin explained this lack of fossil evidence, not as an indication of its non-existence, but simply that it had not yet been discovered, and he expressed the hope that, in future years, his theory would be substantiated by subsequent discoveries.[10] In other words, Darwin was saying, "*I can't substantiate my theory with hard evidence, but one day I hope others will be able to do so.*" This is why Darwin wrote to his colleague, admitting;

> "*I am quite conscious that my speculations run quite beyond the bounds of **true science**.*"[11]

This is an extraordinary admission! In fact, it represents a complete lack of scientific methodology. True science, as Darwin refers to, involves the testing of theories against measurable, repeatable evidence in the natural world. If a theory cannot be replicated and verified, and is not supported by quantifiable, observable evidence, it cannot be regarded as a valid scientific theory. Such a theory must either be modified to fit the observable data or completely discarded. In the next chapter of this book, we will investigate the complete lack of palaeontological (fossil) evidence for Darwin's theory. This is now a major problem for the theory of evolution; one which has brought it to the brink of scientific extinction.

There is a fascinating admission in the Introduction to the 6th edition of Darwin's "*On The Origin of Species*", 1956,[12] written by Dr. W. R Thompson, Entomologist and Director of the Commonwealth Institute of Biological Control, Ottawa, Canada. He wrote;

> "*Darwin himself considered that the idea of evolution is unsatisfactory unless its mechanism can be explained. I agree, but since no one has explained to my satisfaction how evolution could happen I do not feel compelled to say that it has happened. I prefer to say that, on this matter, our investigation is inadequate.*"[13]

In other words, even a scientist who is positively inclined towards Darwin, to the point where he was invited to write the introduction to the 6th Edition, could not bring himself to say that the theory, after nearly 100 years of further research, is supported by hard evidence! Because it is not. Hence, his frank admission of doubt; "*I do not feel compelled to say that it [evolution] has happened*".

How, then, did Darwin come to propose his audacious theory? Surely there was some basis for his speculations? What led him to his conclusions?

Darwin's theory of evolution rests upon an assortment of simple observations:

- Variations of features within the same species of grass and plants.

- The shared basic structure of all flowers (all flowers have petals, sepals, stamens and pistils).

- Variations of beak length in finches on the Galapagos Archipelago.

- The discovery, on the archipelago of Madeira, of wingless beetles living alongside winged beetles.

- The similarity in appearance of the early stages of embryos of almost all species.

- Similarities in physical structures among widely different species. (For example the similarity in limbs between many different species).

That was it! That was the basis of Darwin's theory that fish turned into mice, dogs, elephants, birds and humans!

In regard to Darwin's observations of variations **within** species, this is recognised today as a simple result of genetic specialisation, sometimes referred to as micro-evolution. There is no doubt that changes can occur *within* a species over time, resulting in many sub-species. For example, the canine species we refer to as "dogs" currently has about 340 sub-species or breeds, all of them with different physical characteristics; big, little, long-haired, short-haired, fat, skinny, fast, slow, etc. Genetically, all these breeds originated from a single wolf-like canine, and have diversified and specialised over the centuries. The gradual development of different breeds is the result of dogs with similar characteristics breeding with each other and strengthening those characteristics within each subsequent generation. The result is that the genetic information for the weaker traits is eventually lost. For example, if dogs with slightly longer hair breed with similar dogs, eventually (after many generations) the genetic information for short hair is completely lost, while the genetic information for long hair is strengthened, and a "new" breed is created which is only capable of producing long haired offspring.

This process of micro-evolution within a species is a well-documented genetic fact, resulting from the specialisation and *loss* of genetic information. However, it is quite impossible for a dog to eventually become an elephant or an eagle. For this to occur, the dog would need to acquire *new* genetic information (to grow a trunk or to grow wings). As we will see in a subsequent chapter, this is simply impossible. There is no possible natural process by which new genetic information can be introduced to the replication process. Genetic mutations only ever result in the degradation or complete loss of genetic information, not the addition of new information.

In regard to Darwin's observation of the physical similarity of embryos across many species, Darwin made the extraordinary leap of logic to conclude that the reason for this similarity must be that we all originated from the same original life-form. For example, almost all animal embryos, in their early stages, seem to have a tail and gill-like structures. Therefore, Darwin and his contemporaries concluded that this is evidence that we all originally descended from fish! In his "On The Origin of Species", he writes;

> *"The embryonic state of each species and group of species partially shows us the structure of their less modified ancient progenitors."*[14]

This leap of logic persisted for over 100 years, with highly inaccurate, exaggerated drawings of fish-like embryos of land animals appearing in science text books right up until the 1990s. Biologists have now completely debunked this imaginative hypothesis, by proving that the embryonic "tail" of land animals is simply the developing spinal chord, and bears no similarity to the skeletal and muscular characteristics of embryonic fish tails. Biologists have also completely repudiated the idea that there are gill-like structures in embryonic land species; these were a fabrication initiated by highly stylised drawings by German Zoologist, Ernst Haeckel, a contemporary of Charles Darwin.[15]

In regard to Darwin's observation of the similarity in the physical structures of various species, a similar leap of logic took place. Darwin assumed, for example, that because many land animals had the same basic structure of four limbs with similar articulating joints, the only explanation of this is that they all had the same common progenitor; that mice and elephants are related to each other because they both have four legs with two joints! This is akin to saying that because all cars have four wheels and a steering wheel they must have evolved of their own accord from a single original car. This is simply illogical. The reason why different model cars all have similar structures is because the designers have found that these structures are the most efficient, and so the same structures are used in the production of all makes and models.

In the case of similarities of structure between different animal species, a much more logical conclusion is that the Creator designed structures that worked efficiently, and chose to utilise those structures repeatedly in the creation of all the different species.

# Darwin's Imaginative Theory

**WHAT DARWIN DID NOT KNOW**

Charles Darwin lived at a time when the various branches of science were still in their infancy, and many crucial discoveries had not yet taken place. The science of the 1800s was ignorant of many of the simple facts that we take for granted today, and was encumbered by many illogical and ill-founded beliefs. For example, until the mid-19th century, some scientists believed that it was possible to turn lead into gold, many more scientists believed that rotting vegetation spontaneously generated living organisms including rats and mice![16]

Darwin's theory was severely hindered by his ignorance of scientific facts that were not yet known:

**DNA Discovered in 1953**

This is the genetic building building-block of life, containing, in the case of humans, the encoding for the construction and proper functioning of the entire human body. This complex double-helix chain of deoxyribonucleic acid (DNA) is found in almost every cell of the body and contains 3.2 billion pieces of information. It is a highly complex information system that must have been present from the very beginning of life, and evolution has no known means of accounting for its creation. Where did this huge amount of genetic information come from? Evolution has no answer.

**Molecular Genetics and Genetic Replication**

The discovery of DNA in 1953 heralded the beginning of our understanding of how genetic reproduction takes place. Darwin's theory requires the addition of huge amounts of new genetic material, from purely random processes, for the evolution from simple to complex species to occur. The modern science of genetics now reveals that genetic mutation during cell reproduction *never* adds new meaningful genetic information. Rather, it results in genetic information being either degraded or lost entirely. Charles Darwin, of course, had no understanding of this when he proposed his theory.

**Irreducible Complexity of Cells**

Cells were once thought to be simple building blocks of life, but we now understand that this is not so. Cells are incredibly complex. Even the simplest cell is full of hundreds of molecular machines, each of which is comprised of dozens of independent parts, formed by the construction of DNA and RNA chains, each of which was constructed by other molecular machines inside the cell. And all of these machines have to exist simultaneously in order for a single cell to function and remain alive. The simplest of cells is unbelievably complex! Evolutionary theory has no satisfactory way of explaining how hundreds of these micro-molecular machines could spring into existence simultaneously in order to create the first living cell.

Charles Darwin once stated;

> *"If it could be demonstrated that any complex organ existed which could not possibly have been formed by numerous, successive slight modifications, my theory would absolutely break down."*[17]

Commenting on Darwin's statement, and in the light of our recent knowledge of the irreducible complexity of cells, Dr. Michael Behe, Professor of Biochemistry at Lehigh University in Pennsylvania and a senior fellow of the Discovery Institute's Centre for Science and Culture, states;

> *"As the number of unexplained, irreducibly complex biological systems increases, our confidence that Darwin's 'criterion of failure' has been met skyrockets towards the maximum that science allows."*[18]

# Darwin's Imaginative Theory

If Charles Darwin was alive today, one wonders whether he would concede that his theory, according to his own defined parameters, has been refuted by overwhelming scientific evidence.

## Lack of Fossil Evidence

Although Darwin failed to find any fossil evidence to support his theory, he lived in hope that subsequent generations of scientists would do so. His belief was that if evolution had occurred as he theorised, there should be vast numbers of transitional forms lying in the ground, awaiting discovery. Darwin was correct in one sense. If evolution is true, there **should** be vast numbers of transitional forms providing incontestable proof of the theory. The fact remains, as we shall see in succeeding chapters, that there is not one such incontestable fossil in existence today. Darwin went to his grave hoping for proof that has never eventuated. As Prof. E.H. Andrews states;

> *"The fossil record now constitutes a severe embarrassment to the theory of evolution!"*[19]

## THE POPULARITY OF DARWIN'S THEORY

If there was such pitiful evidence for Darwin's theory when he first proposed it, how did it gain such traction within the scientific community? And why is it still so stubbornly upheld today, despite the growing contradictory scientific evidence? The answer is simple; because it provides atheists with an explanation for the origin of life that does away with the need for a Creator. The theory of evolution has been passionately embraced by a large section of the scientific community because it provides them with an apparently scientific basis upon which to rest their atheism. As I quoted in the previous chapter, Richard Dawkins stated;

> *"Darwin made it possible to be an intellectually fulfilled atheist."*[20]

This is why evolutionists will fight ferociously to defend the theory, despite overwhelming and growing contradictory evidence, because, at its heart, evolution is a philosophy, not a science. In case you think I am being overly dramatic at this point, this is also the conclusion of the French scientific organisation, CESHE (Cercle d'études Scientifique et Historique). This is an international organisation of geneticists and other scientists, formed in 2002. One of their purposes was to investigate whether, given our current scientific knowledge, the theory of evolution can still be considered a viable scientific theory. After many months of investigation and rigorous evaluation of all the latest evidence, this was their published conclusion;

> "*The theory of evolution is not supported by science.* Many scientists have accepted the theory because they assume it to be an established scientific fact. Those scientists who have investigated it, however, find that *evolution is a belief, not a science.*"[21]

---

## ENDNOTES - Chapter 2

---

[1] From a letter to Asa Gray, Harvard biology professor, cited in "Charles Darwin and the Problem of Creation", N.C. Gillespie, p.2

[2] Cited in "*Darwin's Illegitimate Brainchild*", by Russell Grigg, on Creation.com website

[3] Charles Darwin, "On The Origin of Species", London, John Murray, 1859, p.156

[4] Graham Lawton, "New Scientist", 24th Jan, 2009, p.34

[5] https://creation.com/exploring-evolution-darwinism

[6] Darwin online, *On The Origin of Species*, pp. 70 ff., 6th edition, 1872.

[7] Charles Darwin, *On The Origin of Species*, 6th edition, 1872

[8] IBID, Chapter 6

[9] IBID

[10] IBID

[11] From a letter to Asa Gray, Harvard biology professor, cited in *"Charles Darwin and the Problem of Creation"*, N.C. Gillespie, 1982, p.2

[12] Introduction to "The Origin of Species", 6th Edition, 1956, p. 25

[13] Introduction to "The Origin of Species", 6th Edition, 1956, p. 25

[14] Darwin, "On The Origin of Species", op. cit. p.427

[15] Cited in "Evolution Impossible", Dr. John Ashton, Master Books, Green Forest, AR, 2013, p.33

[16] Matt Simon, "Fantastically Wrong: Why People Once Thought Mice Grew Out Of Weat And Sweaty Shirts", https://www.wired.com/2014/06/fantastically-wrong-how-to-grow-a-mouse-out-of-wheat-and-sweaty-shirts/

[17] Charles Darwin, "On The Origin of The Species", 1st edition, 1859, p.189

[18] Michael Behe, "Darwin's Black Box; The Biochemical Challenge To Evolution", 1996, p.39

[19] Quoted in *"Evolution; The Lie"*, Ken Ham, New Leaf Press, 2007.

[20] Dawkins R., *The Blind Watchmaker*, Penguin, London, p. 6, 1991

[21] cercle d'études scientifique et historique, https://ceshe.fr/revue-sf-1.html#p=3, also quoted in "Evolution; Fact or Belief?", Creation Science Foundation.

The Problem with Fossils

Chapter 3

# The Problem with Fossils

Charles Darwin's hope that fossil evidence would eventually be discovered to support his theory was utterly misplaced. Not only has no unequivocal corroborative fossil evidence been discovered, but all the evidence that we *do* have points very strongly to sudden creation and, as will be explained, a sudden catastrophic inundation that resulted in the formation and preservation of fossils all over the earth.

**THE ABSENCE OF UNCONTESTED TRANSITIONAL FORMS**

If evolution is true - if all species gradually evolved over millions of years through long, slow stages of increasingly complex transitional forms (intermediate species) - the earth should be overflowing with fossils demonstrating this. The rocks and soil should be teeming with transitional forms of all the various stages between emerging species. We should be able to find abundant fossil evidence of transitional forms between reptiles and birds, primates and humans, and all the various species that have supposedly evolved. There should be millions of these fossils all over the earth! Yet the fact remains that, after centuries of archaeological excavation, not one single incontestable transitional form has been uncovered.

In terms of supposed human evolution, the modern theory of human evolution proposes that humans evolved through an ape-like ancestor, which is shared by modern apes and monkeys. This gradual emergence of humans from the animal kingdom is supposed to have taken place over hundreds of thousands of years, with vast numbers of incrementally humanoid creatures roaming the earth and leaving behind clear evidence of their existence. The fossil record should be replete with the record of their sojourn upon the earth. But it is not. Not a single such unequivocal and uncontested pre-human transitional form has been discovered.

This is not for lack of trying, however! During the 20th century, over-enthusiastic evolutionists proposed a number of such "missing links". These supposed primate-human transitional forms were announced with great fanfare and were broadcast enthusiastically to the general public as verifiable proof of Darwin's theory. For several decades, discovery after discovery was announced; Java Man, Piltdown Man, Peking Man, Neanderthal Man. These and other "missing links" found their way into science textbooks around the world and were taught in schools and universities as indisputable facts. Several generations of people grew up with the names of these transitional forms in their heads and on their tongues. Evolution had been proven! Darwin had been vindicated!

Not so. Every one of these so-called "missing links" has now been completely discredited. They have all been proven to be either hoaxes by fanatical evolutionists or false hopes based upon very bad science. Here are some examples:

# The Problem with Fossils

- **Hesperopithecus** was believed to be one such missing link, but in the late 1960s, Henry Fairfield Osborn's field expedition proved beyond doubt that Hesperopithecus was the remains of a modern-day wild pig![1]

- **Java Man** was discovered in 1891 by Eugene Dubois. Yet Dubois also discovered fully human skulls at the same level as Java Man, and concealed them for 30 years. Before he died, he confessed this omission of facts, and admitted that Java Man was really a gibbon! Furthermore, Frau Selenka's expedition, in 1907, discovered that the Java Man site was a volcanic area and could not be more than 5000 years old.[2]

- **Piltdown Man**, discovered in 1912 by Charles Dawson, was chemically analysed in 1953 by Prof. Kenneth Oakley, who proved conclusively that the skull was that of a modern human and the jawbone was that of an ape. The bones had been chemically treated by Dawson to make them appear old, and the teeth had been filed down to resemble human teeth. Charles Dawson was disgraced by the eventual unveiling of this fraud, which had fooled the scientific world for 40 years.[3]

- **Neanderthal Man**, discovered in 1848 at Forbes Quarry, Gibraltar, was declared by evolutionists at the time to be THE missing link. In 1939, it was proved, by Prof. Sergio Sergi, that Neanderthal Man had walked erect, and not on all fours as evolutionists had previously believed. Then, in 1947, a Neanderthal Man was discovered to have lived in a cave AFTER modern man had inhabited it. Neanderthal Man is now conceded to have been simply a normal variation of modern humans.[4]

- **Australopithecus Ramidus** was hailed, in 1994, to be an indisputable primate-human transitional form. Scientists have now unearthed a nearly completed skeleton of the same creature and have had to re-classify it as Ardipithecus Ramidus - a modern day monkey.[5]

- **"Lucy"** (an Australopithecine) is the latest supposed missing link hailed by evolutionists. However, living specimens of this creature have been discovered in the jungles of Sumatra. The creature, known as Orang Pendek, is simply another variety of the Orang monkey species, and not, as we were originally told, an ancient ancestor of homo sapiens. The well-known French science journal, "Science et Vie", admitted this fact in its February 1999 issue with the headline, "*Adieu Lucy*" (Farewell Lucy), and published the clear statement that Lucy could no longer be considered an ancestor of modern humans.[6]

The Smithsonian National Museum of Natural History in Washington DC is the premier museum in America, and is colloquially considered to be the "head office" of evolution. It used to have an exhibit called *"Origin Of Life: Apes to Man"*, which featured impressive representations of the various transitional ape-man forms. As a result of the discrediting of all supposed pre-human transitional forms, the display was eventually closed down in the 1980s, and a sign was placed outside it, with the announcement:

> *"A lot has happened since this exhibit opened in 1974. The science of human evolution is a fast-changing field. Much of the material here is now out of date. We are developing a new exhibit based on the latest findings."*[7]

In 2010, the museum opened a new display, *"The Hall of Human Origins"*, which features a progression of skulls supposedly demonstrating evolutionary development over time.

# The Problem with Fossils

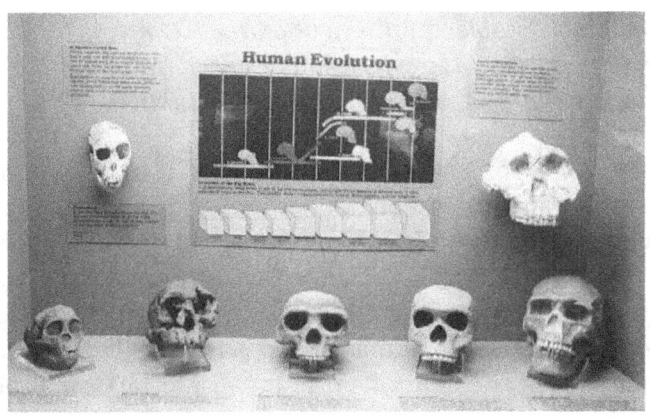

There are two very disappointing things about this exhibit:

**Firstly**, the display features Neanderthal man, homo-habilis and homo-erectus as supposedly pre-human and early-human transitional forms, yet many palaeontologists regard these as being normal variations of modern humans.[8] Dr. Malcolm Bowden, in his article, *"Homo-erectus - A Fabricated Class of Ape Men"*, comments:

> *"ALL skulls can be identified as being either ape or [modern] human. There are NO other classes, for they are all the imaginings of the evolutionary palaeo-anthropologists who insist on concocting a string of links between man and apes."*[9]

**Secondly**, the current evolutionary method of classifying skulls as being pre-human is completely superficial and arbitrary. The skulls in the display that are classified by evolutionists as being transitional pre-human or early-human forms (such as homo-habilis and homo-erectus) are assigned those categorisations based solely upon their *appearance*. Key characteristics for such classification include prominent brow ridges, flat or receding forehead, thick skull and a long, low vaulted cranium. Yet, as Dr. Peter Line explains, in his article, "Fossil Evidence Against Alleged Ape-Men":

# The Problem with Fossils

> *"A major problem for evolutionists is that all of the above-mentioned features, which supposedly differentiate erectus from modern humans, also occur in modern humans."*[10]

Many palaeontologists such as Dr. Line argue that modern human skulls display a wide range of shapes and sizes, and that the evolutionary classification system based upon skull appearance is completely without scientific basis. Dr. Line provides numerous examples of modern human skulls which have been uncovered that display exactly the same features as skulls of supposedly pre-humans. The picture below, published by Dr Line,[11] is the skull of a person who died in the late 1800s in Broken Hill.

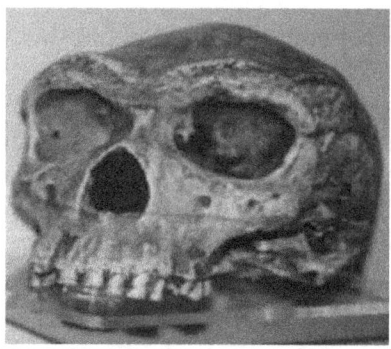

Note the prominent brow ridges, sloping forehead and flat skull. If this skull had been uncovered in an archaeological dig elsewhere in the world, it would undoubtedly have been classified as a pre-human transitional form. Yet it is the skull of a modern human who lived in the 1800s!

Compared to this skull, consider the two skulls below. The one on the left is a Neanderthal (supposedly 400,000 years old) and the one on the right is a Homo-habilis (supposedly 2 million years old). Compared to these two supposedly ancient skulls, the Australian skull, above, appears even more ancient, with much more prominent brow ridges and a flatter forehead and skull!

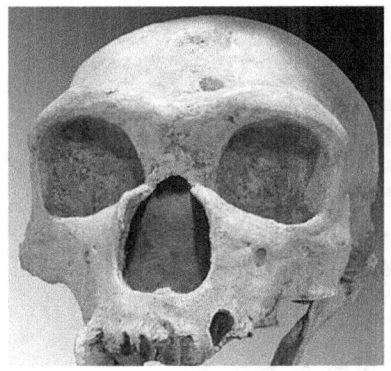 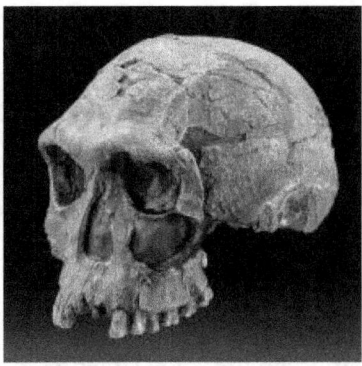

Neanderthal and Homo-habilis

The evolutionary classification of skulls, based on morphological characteristics (supposed changes in appearance over time) is, quite simply, farcical. Dr. Stephen Molnar, in his book, "*Races, Types and Ethnic Groups*", documents size variations in present-day human skulls of up to 2,200 cm$^3$, and also provides evidence of all the various skull shapes and characteristics that are usually attributed to ancient pre-humans as still being prevalent within the human population today.[12]

## NO UNCONTESTED TRANSITIONAL FORMS FOR *ANY* SPECIES

The result, in recent years, of more rigorous examination of archaeological evidence, has been the repudiation of *all* previously held transitional forms. There remains today not one single, uncontested transitional form as evidence for evolution between *any* species; fish to amphibians, amphibians to reptiles, reptiles to birds.

For example, Archaeopteryx was, for decades, widely regarded as the quintessential transitional form between reptiles and birds. As in the case of the supposed ape-human transitional forms, it found its way into science textbooks around the world and was taught to generations of students as archaeological proof of evolution. More thorough investigation in recent decades, however, has totally discredited

Archaeopteryx as a transitional form. It is now known to be a normal variation of a modern bird.

Commenting on the demise of Archaeopteryx as a transitional form, Dr. Stanley A. Rice, in his highly regarded "Encyclopedia of Evolution", states;

> *"Modern birds are not the descendants of Archaeopteryx, which has proved to be an evolutionary dead end. Birds diversified well before Archaeopteryx."[13]*

Similarly, the Berkeley University website states;

> *"It has long been accepted that Archaeopteryx was a transitional form between birds and reptiles. However, its feathers, wings, furcula ("wishbone") and reduced fingers are all now acknowledged as being characteristics of some modern birds."[14]*

In 1999, Colin Patterson, one of the world's leading evolutionary palaeontologists, based at the British Museum of Natural History, wrote the landmark book, *"Evolution"*. In it he failed to mention a SINGLE evolutionary transitional form. When a Christian scientist, Luther D. Sunderland, wrote and asked why he had failed to mention any transitional forms, Patterson wrote back, saying,

> *"I fully agree with your comments on the lack of evidence of evolutionary transitional forms. If I knew of any (either fossil or living) I would certainly have included them in my book. I'll lay it on the line – there is NOT ONE SUCH FOSSIL [emphasis mine] for which there is a watertight argument."[15]*

This now famous quote has severely embarrassed evolutionists, and Patterson, under pressure from the atheist movement, has subsequently tried to qualify his original comment.

If evolution was true, we should expect to find MILLIONS of transitional forms all over the earth. Yet there is not one such transitional fossil! As

# The Problem with Fossils

Prof. E.H. Andrews states, *"The fossil record now constitutes a severe embarrassment to the theory of evolution."*[16]

Dr. Roberto Fondi, Professor of Palaeontology at the University of Seaella, Italy comments;

> *"There remains today not ONE SINGLE transitional link between any two species. The theory of evolution is not supported by palaeontology."*[17]

In this sense, nothing has changed since Darwin first proposed his theory. The absence of transitional intermediaries was deeply troubling even to Darwin's loyal supporters, such as T. H. Huxley, who repeatedly warned Darwin in private that his theory did not match the evidence.[18] Even Darwin himself was perplexed by the complete absence of transitional forms. Commenting on this his *"On The Origin of Species"*, he wrote;

> *"As, by this theory, innumerable transitional forms must have existed, why do we not find them embedded in countless numbers in the earth's crust? ... Why, if species have descended from other species by insensible fine gradations, do we not everywhere see innumerable transitional forms? Why is not all nature in confusion instead of the species being, as we see them, well defined [in the fossil record]."*[19]

In a further admission, Darwin states;

> *"The distinctiveness of specific forms and their not being blended together in innumerable transitional links is a very obvious difficulty."*[20]

The original explanation by Darwin and his contemporaries for this perplexing absence of transitional forms was that there were still undiscovered gaps in the fossil record. According to this explanation, the problem lay not in the theory itself, but in our limited discovery of the fossil record. Darwin refers to this explanation when he states;

## The Problem with Fossils

> "We have seen that whole groups of species sometimes falsely appear to have abruptly developed; and I have attempted to give an explanation of this fact, which if true <u>would be fatal to my views</u>."[21]

This convenient explanation must have had a hollow ring to it, even as it was proposed. What are the chances that palaeontologists failed to find ANY single example of ALL the transitional forms between ALL the stages of development of ALL the species? What kind of statistically improbable bad lack has led palaeontologists to miss the millions of transitional fossils around the world, and to only find fossils of fully developed species?

After a century and a half of further exploration of the fossil record, this enormous gap in the fossil record persists. Modern palaeontologists are reaching the conclusion that the fossil record is complete as it stands, and that, if evolution is true, some other explanation is required. Dr. E.C. Olson states;

> "Many new groups of plants and animals suddenly appear, apparently without any close ancestors. This aspect of the record is real, not merely the result of faulty or biased collecting. A satisfactory theory of evolution must take this into consideration and provide an explanation."[22]

Surely the most logical explanation, however, for the complete absence of transitional forms in the fossil record is that they never existed in the first place and that evolution did not occur. How can science persist in filling the evolutionary gaps with imaginary animals when there is not the slightest physical evidence for their existence?

### TRANSITIONAL FORMS WOULD BE WEAKER

There is also a simple logical argument that needs to be considered here. Any supposed transitional forms between individual species would have been weaker, not stronger, and would not have been an advantageous evolutionary development. The theory of evolution proposes that gradual successive mutations gave the resulting offspring an evolutionary advantage over offspring that did not have those mutations, so that the mutated forms proliferated while the original forms died out. In this way, it is believed that fish gradually changed into amphibians, amphibians changed into reptiles, and reptiles gradually changed into birds and mammals.

Yet simple logic tells us that these imaginary transitional forms would be at a distinct evolutionary *disadvantage*. A fish with half formed legs instead of fins would not be able to swim as efficiently and would be easy prey for predators. A reptile with half formed wings instead of arms would not be able to fly and would also be less able to use its arms or front legs for survival. Wings only give a creature an evolutionary advantage when they are fully formed. Before then, they are a distinct disadvantage and would result in the transitional form being considerably weaker that its non-mutated counterparts, and less likely to survive.

**TEXTBOOK DECEPTION**

Over the decades since Darwin first proposed his theory, scientists have been very quick to hail new scientific "proofs" of evolution and publish the evidence in textbooks, and extremely reticent to retract them when they have proven false. For example, when I graduated from High School in 1977 at the age of 18, science was my favourite subject. Looking back on what I was taught, however, I find it extraordinary that I was still being taught the ape-human "missing links" as being indisputable proof of evolution. Despite the fact that by the mid-1950s all the supposed missing links, up to that point, had been completely discredited, they were *still* in my science textbooks in the mid-1970s, and I still had to regurgitate them in exams as proof of evolution!

## The Problem with Fossils

A more recent example further illustrates this reticence. From 2002 to 2017, I taught Christian Studies in a Christian High School. Towards the end of that time, in 2016, I became curious regarding the content of our current science textbooks. What evidence was being promoted as proof of evolution to our impressionable teenagers? I decided to investigate, so I borrowed the senior biology textbook from the school library; "HSC Biology, 3rd Edition",[23] by Judith Brotherton and Kate Mudie. I was absolutely gob-smacked by what I discovered!

In Chapter 2.1, *"Evidence For Evolution"*, 18 out of the 20 pages discussed observable changes within a species as proof of evolution. Yet even the most uninformed biologist knows that micro-evolution (variations within a species) cannot give rise to macro-evolution (one species changing into a completely new species). This is because of the universally acknowledged impossibility of genetic mutations creating the vast amounts of new genetic information necessary for a new species with brand new physical characteristics to develop. In fact, the impossibility of micro-evolution to give rise to macro-evolution was the central topic of the Chicago Conference of 1980, a meeting of the world's leading geneticists. After considering all the currently available evidence, the conference issued a definitive declaration;

> *"The central question of the Chicago conference was whether the mechanisms underlying micro-evolution can be extrapolated to explain the supposed phenomena of macro-evolution. At the risk of doing violence to the opinions of some of the scientists at the meeting, the answer was a clear 'No'."*[24]

Yet, 36 years after the world's leading scientists had almost universally declared that micro-evolution was **not** evidence of the viability of macro-evolution, it was still the major justification for evolution in our current High School biology textbooks!

The remaining 2 pages of the 20-page chapter, *"Evidence For Evolution"*, proposed 4 transitional forms:

# The Problem with Fossils

- **Crossopterygian:** A fish with a detailed bone structure in its dorsal fins, supposedly the precursor to amphibian limbs. Only a few ardent evolutionists briefly proposed this fish as a transitional form. This is because; (a) its bones are not joined to its backbone as are amphibian limbs, and (b) there are still living specimens of fish with similar bones. No serious palaeontologist still proposes this as a transitional pre-amphibian form.

- **Therapsid:** An extinct group of dinosaurs briefly proposed by a few evolutionists as reptile-mammal transitionary forms. It, too, was quickly discounted because; (a) there is no clear chronological sequence in the fossil record indicating these as transitional forms, (b) the difference in jaw shape is simply normal variation within species, and (c) there are no transitional forms between warm and cold-blooded species.

- **Archaeopteryx:** An extinct bird with claw like bones at the extremities of its wings, once proposed as a transitional reptile-bird form. It has been completely discounted by palaeontologists for over a decade, as previously explained.

- **Darwinius Masillae ("Ida"):** A small, long-tailed monkey, once proposed as a transitional form between different subspecies of primates. It has now been completely discounted, and the previous claims have attracted intense criticism because of the poor science involved. Even the populous Wikipedia admits this:

*"Concerns have been raised about the claims made about the fossil's relative importance and the publicising of the fossil before adequate information was available for scrutiny by the academic community. Some leading biologists, among them Nils Christian Stenseth, have called the fossil an "exaggerated hoax" and stated*

*that its presentation and popular dissemination "fundamentally violate scientific principles and ethics."[25]*

That is all the evidence the HSC Biology textbook can offer! Four transitional forms that have been discredited and discarded by the scientific community! It is obvious what lies behind this intransigence, this dogmatic refusal to relinquish outdated evidence, and in one sense it is understandable. Because, if the authors of *"HSC Biology, 3rd Edition"* deleted from their chapter all the evidence that has been discredited by the wider scientific community, the chapter would be completely blank!

After my disturbing experience with the HSC Biology textbook, I revisited my school library to borrow the Year 10 Science textbook, *"Oxford Insight Science, Stage 5"*. Chapter 2.3, *"Evidence for Evolution"*, only offered one transitional form, Archaeopteryx. Then, on page 89, it made this gob-smackingly ridiculous statement:

> *"It is thought that life originated in the sea, crawled onto land and then took to the skies. But what evidence links these stages? Transitional fossils show intermediate states between the "before" and "after" stages. They are sometimes referred to as missing links, but knowledge of geology and evolution allows palaeontologists to <u>predict the location of transitional fossils and find them.</u>"[26]*

Really? What an extraordinary statement! *"Dennis, be so kind as to pass me the shovel please. I have a feeling in my waters that there is a transitional form under this rock."* How utterly ridiculous! If this claim in the science textbook is true, why haven't we found **any** such fossils? This is not merely a slight exaggeration; it is an example of deliberate deception.

The chapter concludes with the final statement;

> *"When Darwin first published his theory of evolution, he stressed that the lack of transitional forms was the most formidable*

> *obstacle to his theory... Since then, numerous examples have been found, starting with the discovery of Archaeopteryx. All agree it is an important transitional species.*"[27]

Numerous examples? Really? I suppose this statement is correct in one sense, because **zero** is a number! Statements such as these do not deserve to be in a science textbook. Science is about examining the real evidence and critically evaluating and refining theories until they explain that evidence. Instead of this, however, this textbook presents Year 10 students with exaggerations, embellishments and blatant lies.

Sadly, these two textbooks are indicative of the desperation and deception that characterises the way in which educational institutions continue to foist bad science upon our next generation in order to prop up a failing and increasingly discredited theory.

## PUNCTUATED EQUILIBRIUM

So, what is the *actual* fossil evidence? Instead of gradual transition from one species to another, the *actual* evidence of the fossil record shows all the major species appearing suddenly and fully formed. And instead of change over time, the fossil record shows stasis; species remaining basically unchanged to the present day. This sudden appearance of fully-formed species in the fossil record, together with long-term stasis, completely contradicts evolutionary theory.

A landmark study of the fossil record was published in 1972 by Drs. Stephen Gould and Niles Eldredge, entitled *"Punctuated Equilibria"*.[28] These eminent scientists were not creationists; in fact they were atheists and evolutionists. They concluded that the gradualism predicted by Charles Darwin (gradual evolution of species through innumerable, small, successive variations) is virtually non-existent within the fossil record. Instead, they indicated that the reality is one of stasis (species remaining the same throughout time), punctuated by the sudden appearance of fully-formed, completely new species with no intermediaries.[29]

In the words of Gould;

> *"The history of most fossil species includes two features particularly inconsistent with gradualism [evolution]:*
>
> ***1. Stasis.*** *Most species exhibit no directional change during their tenure on earth. They appear in the fossil record looking pretty much the same as when they disappear; morphological change is usually limited and directionless.*
>
> ***2. Sudden appearance.*** *In any local area, a species does not arise gradually by the steady transformation of its ancestors; it appears all at once and fully formed."*[30]

The importance of Gould's and Eldredge's findings cannot be overstated. Their conclusions continue, to the present day, to pose profound challenges to secular evolutionary scholarship.

Subsequent to the publication of the *"Punctuated Equilibria"* study, further studies of stasis in the fossil record have corroborated its findings. Dr. Carl Werner's book, *"Evolution: The Grand Experiment, Vol 2 - Living Fossils"*, published in 2009,[31] provided photographic evidence and hard scientific analysis documenting the fact that the fossil specimens of organisms, supposedly millions of years old, that have survived through to the present time are essentially unchanged. The title of the book, *"Living Fossils"* indicates the nature of Dr. Werner's findings; that there has been no appreciable evolution of surviving species of ancient organisms, despite the passage of, in some cases, supposedly hundreds of millions of years - certainly more than enough time for significant evolution to have taken place.

How do evolutionists explain punctuated equilibrium (stasis and sudden appearance with no transitional forms)? As previously explained, Darwin originally suggested that the problem of sudden appearance was simply a deficiency in our discoveries. In other words, according to Charles Darwin, the gap in the fossil record is not a gap in *reality*, but simply a gap in our *discoveries*. Darwin's hope was that subsequent discoveries would soon rectify this problem.

After more than a century and a half of further exploration, however, this explanation is no longer tenable. The sudden appearance in the fossil record of highly developed life-forms persists. As Dr. Phillip E. Johnson points out;

> *"The fossil record today, on the whole, looks very much as it did in 1859, despite the fact that an enormous amount of fossil hunting has gone on in the intervening years."*[32]

The vast majority of Neo-Darwinists (modern proponents of Darwinian evolution) have, therefore, had to discard Darwin's convenient explanation of the gaps in the fossil record. How, then, do they explain punctuated equilibrium? How do they account for sudden appearance followed by stasis in the fossil record?

Many evolutionists propose that the enormous gaps in the fossil record are due to some, as yet, "inexplicable phenomena" in the past which effectively wiped out these transitional forms, along with the accompanying rock strata, from the fossil record. A second explanation is the concept of "mosaic evolution". This theory postulates that the soft body parts of animals may have been evolving while the skeletal structures (which were fossilised) remained static.[33]

In evaluating these explanations, retired Berkeley Professor, Phillip E. Johnson, in his book, *"Darwin on Trial"*, writes;

> *"No doubt a certain amount of evolution could have occurred in such a way that it left no trace in the fossil record, but at some point, we need more than ingenious excuses to fill the gaps."*[34]

One has to wonder how much longer evolutionists can maintain their theory in the face of a complete lack of corroborative evidence and the proliferation of such problematic, contradictory evidence. Their increasingly fanciful explanations do not arise from hard, evidential science but, rather, from imaginative speculation.

## THE CAMBRIAN EXPLOSION

Arguably, the greatest single challenge to evolution posed by the fossil record is the explosion of life in the Cambrian period, supposedly about 540 million years ago. In the pre-Cambrian strata of the geological column, almost all organisms are single celled or symbiotic colonies of single cells. In the Cambrian strata, however, all the major animal groups (phyla) burst onto the scene suddenly (in geological terms) without any trace of the intermediate stages that Darwinian evolution demands.

The highly regarded atheist and evolutionist, Dr. Richard Dawkins, admitted the serious problem that this posed for evolution, stating;

> *"The Cambrian [strata] shows us a substantial number of major animal phyla already in an advanced state of evolution, the very first time they appear. It is as though they were just planted there, without any evolutionary history."*[35]

Dawkins' surprisingly frank admission has been often quoted, and his subsequent attempts to explain the sudden appearance of thousands of fully developed lifeforms in the fossil record have been unconvincing.


Charles Darwin, himself, was aware of the Cambrian dilemma, and commented, in "The Origin of Species";

> *"The [Cambrian] case at present must remain inexplicable, and may be truly urged as a valid argument against the views here entertained."*[36]

Darwin was correct; it is a major problem for the theory of evolution. And nothing has changed in the years since. The sudden explosion of life in the Cambrian era remains, to this very day, completely incompatible with the theory of evolution.

## TRUE SCIENCE VS WISHFUL THINKING

Scientific method involves proposing theories and then testing them against observable, measurable evidence. If a theory is not supported by the evidence, the theory must be modified or, if the inconsistencies are irreconcilable, abandoned altogether. If belief in a theory persists despite being contradicted by observable, measurable evidence, one has to wonder whether it is appropriate to continue to call it a *scientific* theory at all. In regard to the manifest fossil evidence collected over centuries, the theory of evolution is clearly an unsubstantiated philosophical belief, rather than a validated scientific theory.

The obstinacy with which evolutionists cling to their belief, despite all evidence to the contrary, is astonishing. More disturbingly, the tendency of evolutionists to hide the gaping holes in the theory from the view of the general public, and even promote the idea that the theory of evolution is overwhelmingly substantiated by the fossil record, is truly deceitful. Outrageous and highly exaggerated claims by evolutionists abound:

Dr. William H. Matthews says;

> *"Fossils provide one of the strongest lines of evidence to support the theory of organic evolution."*[37]

Drs. Twenhofel and Shrock state;

> *"No line of evidence more forcefully and clearly supports the fundamental principle of evolution - descent with accumulative modifications - than that furnished by fossils."*[38]

More recently, in his book, *"The Greatest Show on Earth"*, Dr. Richard Dawkins made the astonishing claim that;

> *"Evolution is an inescapable fact"*[39]

In Chapter 6 of the same book, Dawkins then makes a further astonishing claim;

> *"Actually, we are lucky to have any fossils at all, let alone the massive numbers that we now do have to document evolutionary history—<u>large numbers of which, by any standards, constitute beautiful intermediates.</u>"*[40]

This claim by Dawkins of *"large numbers"* of transitional forms is not substantiated by reference to any specific, unequivocal examples. It is a classic example of the deceptive sweeping generalisations made by evolutionists to promote a skewed perception of evolution as an indisputably proven theory.

Speaking of this collective deceit by the evolutionist movement, particularly in regard to fossil evidence, Dr. Niles Eldredge, a palaeontologist at the American Museum of Natural History, and a staunch atheist, made a startling admission in his landmark 1985 book, *"Time Frames: Rethinking Darwinian Evolution"*;

> *"We have proffered a collective tacit acceptance of the story of gradual adaptive change; a story that strengthened and became even more entrenched as the synthesis took hold. We palaeontologists have said that the history of life supports it, <u>all the while really knowing that it does not.</u>"*[41]

# The Problem with Fossils

**SUMMARY**

The fossil record represents a major problem for evolution. Instead of millions of fossilised transitional forms all over the earth, as the theory would lead us to expect, there are none. Instead of gradual development from the simple to the complex, we find the sudden appearance in the geological strata of fully formed organisms, without any preceding intermediary stages. Instead of continual evolution of species, we find ancient fossils of organisms that have survived for millions of years into the present day, virtually unchanged. All this fossil evidence accumulated over several centuries, rather than corroborating evolution, strongly contradicts it.

Dr Chandra Wickramasinghe, professor emeritus of University College, Cardiff, UK, once stated:

> "Frequent and massive gaps in the fossil record and the absence of transitional forms at the most crucial stages in the development of life show clearly that Darwinism is woefully inadequate to explain the facts."[42]

Dr. Roberto Fondi, Professor of Palaeontology at the University of Seaella, Italy comments;

> "The fundamental assumptions upon which evolution is based are not at all confirmed by palaeontology. All the biological groups, from bacteria to humans, _appear abruptly in the fossil record, without any links connecting them_. If evolution had really happened, the evidence would be in abundance and incontestable. The museums would be overflowing with exhibits clearly documenting the transitions between various biological groups. But the fact is that after nearly two centuries of intense research, there are NO such exhibits."[43]

Prof. E.H. Andrews puts it more simply; _"The fossil record constitutes a severe embarrassment to the theory of evolution"_[44]

## ENDNOTES - Chapter 3

[1] *"Myths & Miracles"* by David C. C. Watson. Distributed by Creation Science Foundation

[2] IBID.

[3] IBID

[4] IBID

[5] https://australianmuseum.net.au/ardipithecus-ramidus

[6] https://en.wikipedia.org/wiki/Talk%3ALucy_(Australopithecus)

[7] Cited by Ken Ham, https://answersingenesis.org/the-word-of-god/the-wrong-way-round/

[8] https://creation.com/australopithecus-and-homo-habilis

[9] Malcolm Bowden, Homo-erectus - Fabricated Class of Ape-Men, Creation Science technical Journal, Vol 3, 1988, pp.152-153

[10] Peter Line, "Fossil evidence for alleged apemen— Part 1: the genus Homo", p.27, on https://creation.com/images/pdfs/tj/j19_1/j19_1_22-32.pdf

[11] IBID

[12] Molnar, S., *Races, Types, and Ethnic Groups*, Prentice-Hall Inc., NJ, p. 57, 1975;

[13] Stanley A. Rice, "Encyclopedia of Evolution", 2007, p.400

[14] http://www.ucmp.berkeley.edu/diapsids/birds/archaeopteryx.html

[15] Luther Sunderland, "Darwin's Enigma", Master Books, 1998, pp.101-102

[16] Quoted in "Evolution; The Lie", Ken Ham, Creation Science Foundation, 2007

[17] Roberto Fondi, "After Darwin; Evolutionary Criticism", 1980, p. 127

[18] Phillip E. Johnson, "Darwin On Trial", Regnery Gateway Publishing, 1991, p.34

[19] Charles Darwin, "The Origin of Species", Chapter 6, http://www.talkorigins.org/faqs/origin/chapter6.html

[20] IBID, Chapter 9, http://www.talkorigins.org/faqs/origin/chapter9.html

[21] Charles Darwin, "The Origin of Species", New York, New American Library, 1958, p.316.

[22] E.C. Olson, "The Evolution of Life", New York, Mentor Books, 1966, 165, p.94

[23] Judith Brotherton and Kate Mudie, "HSC Biology" 3rd Edition, 2010

[24] Roger Lewin, "Science" Journal, Vol. 210(4472), 1980, pp.883-887

[25] https://en.wikipedia.org/wiki/Darwinius (01/03/19)

[26] "Insight Science, Stage 5", Oxford Press, p.89

[27] IBID

[28] https://en.wikipedia.org/wiki/Punctuated_equilibrium

[29] IBID

[30] Phillip E. Johnson, op. cit., p.37

[31] Carl Werner, "Evolution: The Grand Experiment, Vol 2 - Living Fossils", New Leaf Publishing, 2009

[32] Phillip E. Johnson, op. cit.

[33] IBID

[34] IBID

[35] Richard Dawkins, https://www.newsweek.com/excerpt-richard-dawkinss-new-book-evolution-79345

[36] Charles Darwin, op. cit.

[37] William H. Matthew, "Fossils", New York, Barnes and Noble, 1962, p.47

[38] William Twenhofel and Robert Shrock, "Invertebrate Palaeontology", New York, McGraw Hill, 1935, p.23

[39] Richard Dawkins, "The Greatest Show on Earth", London, Bantam Press, 2009, p.18

[40] IBID, Chapter 6, p.145

[41] Niles Eldredge, "Time Frames: The Rethinking of Darwinian Evolution and The Theory of Punctuated Equilibrium", Simon & Schuster, 1985

[42] Chandra Wickramasinghe, Testimony in court case, Arkansas, USA, 1981, quoted in https://www.panspermia.org/chandra.htm

[43] Roberto Fondi, op. cit.

[44] E.H. Andrews, "God, Science and Evolution", Presbyterian and Reformed Publishing Co, 1981

Chapter 4

# Rocks and The Age of The Earth

Evolutionary theory is completely dependent upon an ancient age of the earth. The gradual cumulative mutations theoretically involved in the evolution of single-celled life into the complex lifeforms of today necessarily require literally billions of years. Even the most ardent of evolutionists acknowledge the impossibility of evolution occurring in anything less. It stands to reason, therefore, that establishing an ancient age of the earth is fundamental to the survival of evolutionary theory.

Evolutionists are adamant that the earth is billions of years old; the consensus seems to be approximately 4.53 billion years old, with the wider universe being 14.5 billion years old.

The following table indicates the widely-accepted timescale of the evolution of biological life on earth.[1] The numbers represent millions of years.

# Rocks and the Age of the Earth

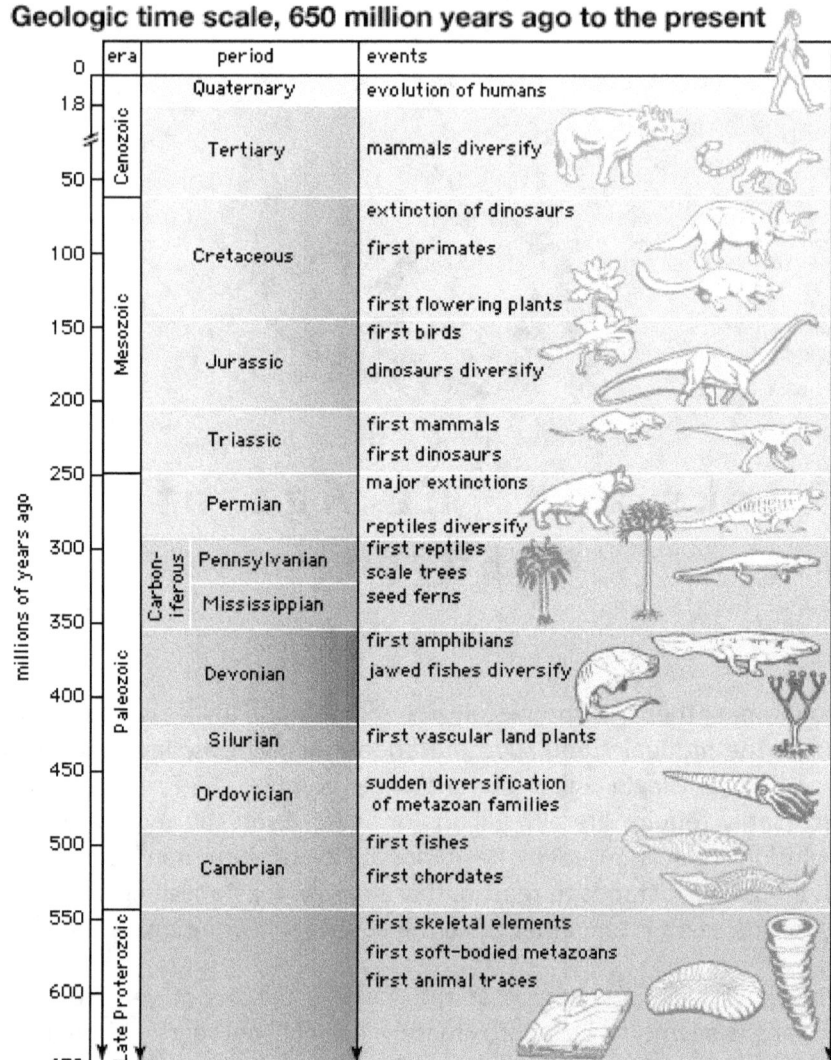

# Rocks and the Age of the Earth

In particular, note the following suggested evolutionary milestones:

- 3.5 billion years ago - Single-celled organisms
- 650 million years ago - Multi-celled organisms
- 500 million years ago - The Cambrian explosion: many invertebrate phyla appear
- 450 million years ago - Fish proliferate
- 400 million years ago - Amphibians and insects appear
- 300 million years ago - First reptiles
- 200 million years ago - dinosaurs dominant, first mammals, and first birds
- 50-60 million years ago - Dinosaurs become extinct
- 55 million years ago - primates evolve
- 1.8 million years ago - Supposed early human transitional forms evolve
- 200,000 years ago - Modern humans (homo sapiens) evolve
- 1971 - Led Zeppelin released their landmark 4th album

Where did evolutionists get these ages from (apart from Led Zeppelin's album)? How did they arrive at these dates? Rocks and fossils do not come with a date stamp. Neither has the universe conveniently printed a chronological historical calendar for us.

Firstly, it must be stressed that no scientific methodology can unequivocally establish the age of the earth. As you are about to see, all dating methods are speculative estimates resulting from calculations based upon a range of unprovable assumptions about the past. Furthermore, as you will also soon see, when the various dating methods have been tested against very recently formed rocks of known ages, they have been shown to be wildly and laughably inaccurate.

There are three main methods that scientists use to date the earth. Relative dating of sedimentary rock layers will be discussed in the next chapter, *"Dirt, Rocks and Water"*. In this current chapter, we will briefly examine the remaining two methods; radiometric dating and Carbon 14 dating, both of which display major inconsistencies.

## RADIOMETRIC DATING OF ROCKS

This is the flagship of evolutionary dating methodology. It is the means by which evolutionists substantiate an old age of the earth (4.53 billion years), which then, it is argued, provides credibility to the whole evolutionary process. Before we examine the over-simplistic and unprovable assumptions that underly this method, together with the huge discrepancies that it throws up, it would be helpful to have a basic understanding of the background science.

### Background Science

Volcanic and igneous rocks (rocks that have cooled from magma, either above or below the earth's surface) contain natural traces of isotopes of various elements. Isotopes are variations of an element with different numbers of neutrons (it doesn't matter if you don't understand this bit). Isotopes may be stable or unstable. Stable isotopes have a stable number of neutrons and protons in their nucleus and, therefore, do not decay. In other words, they will remain the same and will not change over time. Unstable isotopes, also called radioactive isotopes or radioisotopes, have either too many neutrons or too many protons in their nuclei, and try to balance themselves by losing the excess neutrons or protons. (They do this by means of either alpha decay, beta minus decay or beta plus decay

- but you don't need to understand this!). Each time an atom of an unstable isotope loses neutrons or protons, that atom becomes either a slightly different isotope of the parent (original) element or a new element. In this way, an individual atom of an unstable isotope can undergo a succession of cumulative changes, losing neutrons or protons, until it has decayed into a stable form.

For example, Uranium 235 (U-235) decays through a succession of stages until it eventually reaches the stable form of Lead 207 (Pb-207). (The numbers are the total number of neutrons plus protons in the nucleus, referred to as the atomic mass of the isotope). In this case, the decay chain for U-235 to Pb-207 involves 11 stages, as follows:

**U-235 to Th-231 to Pa-231 to AC-227 to Th-227 to Ra-223 to Rn-219 to Po-215 to Pb-211 to Bi-211 to Po-211 to Pb-207**

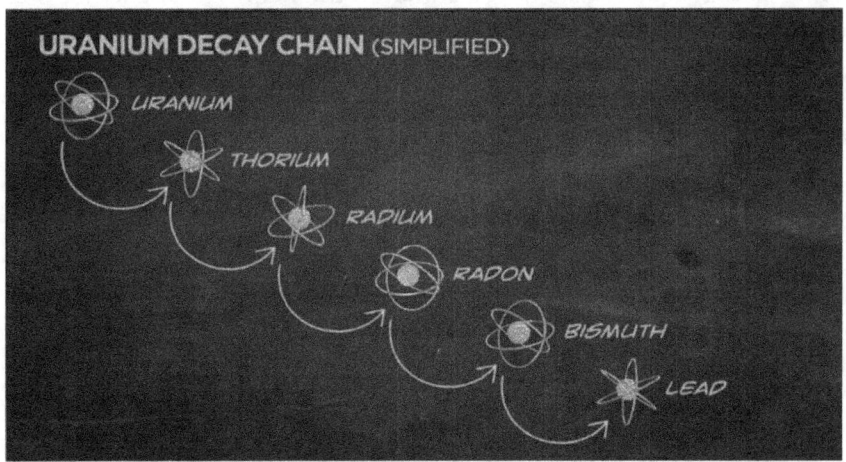

This kind of decay of unstable isotopes into stable isotopes, through a series of intermediate stages, is extremely slow. In the above case, scientists estimate that after 710 million years, half of the U-235 will have decayed into Pb-207. This is referred to as the half-life of an isotope. So, Uranium-235 is said to have a half life of 710 million years.

# Rocks and the Age of the Earth

**The Methodology**

Scientists claim to be able to calculate the age of a rock by examining the amount of original unstable isotope (parent isotope) that it contains, compared to the amount of the resulting stable isotope (daughter isotope). The greater the ratio of daughter isotope to parent isotope within an individual rock, the older it is assumed to be. The determination of a rock's age is made by taking into account the half-life (decay rate) of the unstable parent isotope and calculating how long it would have taken for the amount of daughter isotope that is present in the rock to have been produced by the decay process.

U-235 to Pb-207 is not the only decay reaction that occurs in igneous or volcanic rocks. There are many other naturally occurring unstable isotopes that decay over time, through a succession of changes, into stable isotopes. Scientists have estimated the half-lives of each of these decay reactions.

The following table shows some parent-to-daughter isotopic reactions and half-lives commonly used to calculate the age of rocks:

| ISOTOPIC DECAY REACTIONS | | |
| --- | --- | --- |
| PARENT UNSTABLE ISOTOPE | HALF LIFE | DAUGHTER STABLE ISOTOPE |
| Uranium-235 | 710 million years | Lead-207 |
| Uranium-238 | 4.47 billion years | Lead-206 |
| Samarium-147 | 106 billion years | Neodymium-143 |
| Rubidium-87 | 48.8 billion years | Strontium-87 |
| Rhenium-187 | 43 billion years | Osmium-187 |
| Lutetium-176 | 35.9 billion years | Hafnium-176 |
| Thorium-232 | 14 billion years | Lead-208 |
| Potassium-40 | 1.25 billion years | Argon-40 |

## Rocks and the Age of the Earth

All this sounds like irrefutable science. How could anyone doubt the age of rocks, and the age of the earth itself, given these kinds of highly sophisticated calculations? After all, you can't argue with science!

The problem is that while these kinds of calculations appear to be mathematically certain and scientifically indisputable, they are actually based upon a series of unprovable assumptions and highly speculative guesses. And, as we shall soon see, when each of these dating methods (using different isotopic decay reactions) are compared side by side, they produce wildly varying ages for the same rock.

In particular, radiometric dating of igneous rocks is based upon the following speculative, unprovable assumptions:

1. There were no daughter isotope atoms present in the rock when it was formed. In other words, it is assumed that all the daughter isotope atoms that are currently present within a rock are the result of the decay reaction, and were not already there in the beginning. THE PROBLEM: If naturally occurring daughter isotopes were already present in the rock when it was formed, the rock would have a higher ratio of daughter to parent isotope atoms, thus making it appear much older than it actually is. Even if scientists try to factor into their calculations an allowance for a certain amount of original daughter isotope, this is purely a *guess*, because we have no way of knowing how much daughter isotope the rock would have originally contained.

2. No parent unstable isotope atoms have been leached from the rock over time (or added to it). In other words, it is assumed that the isotopic decay reaction is a closed system, with no outside influences interfering with the reaction. THE PROBLEM: Contact with saltwater is known to leach unstable isotopes from rocks. If some of the parent isotope has been removed in this way, the rock would have a higher ratio of

daughter to parent isotope atoms, thus making it appear much older than it actually is. As it is an established geological fact that most of the land mass of the earth was, at various points, either covered by ocean or subject to seawater inundation, massive isotopic leaching is almost certain to have occurred. Even if scientists try to factor into their calculations an allowance for a certain amount of leaching over time, this is purely a *guess*, as the amount of leaching would vary considerably with different rocks in different locations.

3. The decay rate of unstable isotopes has remained constant over time. THE PROBLEM: Scientists can't go back in time to measure decay rates in the past. They can only measure decay rates in the present. Furthermore, recent research[2] has indicated the strong possibility that decay rates in the past were significantly faster, which would result in rocks appearing to be much older than they actually are.

**Faster Decay Rates in The Past**

If decay rates of isotopes were significantly faster in the past, rocks may be significantly younger than their currently calculated ages. Recent research has indicated a number of factors that could have significantly increased decay rates in the past. For example, a team of scientists, in 2001, were able to substantially alter the half life of Rhenium-187 from the generally assumed 41.6 billion years to a mere 33 years by replicating, in the laboratory, conditions that could plausibly have been present in the primordial past.[3] That is a decay rate *1 billion times faster* than the generally assumed rate!

Recently, scientists at Stanford University discovered that radioactive decay rates fluctuate in synch with the rotation of the sun's core.[4] They also noted changes in the decay rate in response to solar flares and the distance of the earth from the sun, all of which clearly indicates that decay rates are responsive to changes in solar radiation.[5] Although these fluctuations are small, they indicate that the long-held belief that decay rates are constant is not true. Dr. John Ashton argues that changes in cosmic radiation levels and fluctuations in the earth's magnetic field in the past, both of which are highly likely to have occurred, would result in rocks today appearing much older than they actually are.[6]

Significant variations in decay rates have now also been shown to result from variations of temperature[7], pressure and the composition of atmospheric gases[8]. An 8-year study, from 1997 to 2005, called the RATE project (**R**adioisotopes and the **A**ge of **T**he **E**arth) was conducted to asses the reliability of radiometric dating methods. The team of scientists, from the Creation Research Society and the Institute for Creation Research, all of whom are scientists with excellent credentials from major secular universities, and impressive careers in the secular scientific world, published their findings in 2005. They published findings that contradict the traditional view of constant decay rates and indicate that decay rates must have been significantly faster in the past.[9] If this is true, it implies that the actual ages of rocks are much younger than the ages calculated by radiometric dating.

For example, samples of New Mexico granite were recently assigned an age of 1.5 billion years, using the Uranium-Lead method, but the residual helium in the rock which resulted from that decay reaction only indicates an age of about 6,000 years.[10] If this age is correct, it means that the uranium must have decayed at least 250,000 times faster than the currently proposed rate![11]

In his book, "*Evolution Impossible*", Dr. John F Ashton comments;

> "*We have geological evidence of accelerated radioactive decay in the past, which would give the appearance of much greater age to rocks dated by these methods. It includes research scientists' reports on accumulation and diffusion rates from radioactive decay observed in Precambrian granite rocks at Fenton Hill in New Mexico. These suggest that hugely accelerated rates of*

*radioactive decay occurred in the recent past, that is, only thousands of years ago."*[12]

As a result of these recent scientific observations, the uniformitarian assumption that isotopic decay rates were constant in the past is now seriously under question by a small but growing group of concerned scientists. If this fundamental assumption is no longer valid, radiometric dating of rocks cannot be used to establish reliable ages; if decay rates varied in the past, any results arising from calculations based upon current assumptions must surely be regarded as almost meaningless.

**Discrepancies Between the Methods**

If the radiometric formulae for estimating rock ages are accurate, then it should not matter which isotopes in a particular rock are measured, the same age should be indicated. Measuring the parent to daughter ratios of different isotopes in a given rock should give a consistent age, within a reasonable order of accuracy. But this is not what happens. When the same rock is dated using a variety of methods, the different methods *invariably* give widely disparate ages. Not just occasionally, but *constantly*.

Here are a few examples:

- Granite from Ontario was given an age of 390 million years using the Th-232 method, and 1.05 billion years using the U-235 method.[13] This is a difference of 660 million years (or 269%)!

- Granite from Colorado was dated as 313 million years using Th-232, 707 million years using U-235 and 980 million years using the Lead-Lead method.[14] A discrepancy of 667 million years!

- Granite from Arizona was dated as 271 million years using Th-232, 770 million years using U-235 and 1.21 **billion** years using Lead-Lead.[15] A discrepancy of 939 million years!

- Another sample of granite from Colorado was dated as 530 million years using Th-232, 1.13 billion years using U-235 and 1.54 billion years using Lead-Lead.[16] A discrepancy of 1.01 billion years!

- Zircon from Ontario was dated as 390 million years using Th-232, 1.05 billion using U-235 and 1.09 billion years using Lead-Lead.[17] A discrepancy of 700 million years.

- Beolite from Colorado was dated as 1.1 billion years using Th-232, 2.06 billion years using U-235 and 3.18 billion years using U-238.[18] That is a massive discrepancy of 2.08 billion years!

This is just a **very** small sample of the consistently divergent ages assigned by the various dating methods. So how old are these rocks? How can we have any certainty at all? If you had four or five different measuring tapes to measure the length of a piece of timber, and they all gave hugely different lengths, with the largest measurement being **three times as long as the smallest**, how could you reach any level of certainty regarding actual length of the timber? Surely you would not keep using the same tape measures year after year: Surely you would throw them all out and look for a more reliable tape measure!

In a sense, that is precisely what scientists have attempted to do. To try to alleviate these huge anomalies in the different dating methods, scientists have begun using a technique known as isochron dating, in order to smooth out these huge discrepancies. This involves collecting a number of rock samples from different parts of the rock being dated and averaging the parent / daughter isotope readings, while factoring in the parent /daughter isotope ratios of a similar stable isotope. The problem is that isochron dating has demonstrated the same huge inconsistencies. For example:

- Using isochron dating methods, basalt from Somerset Dam in Queensland, Australia was dated as 183 million years using the Potassium-Argon based isochron method, 393 million years using Rubidium-Strontium based isochron method and 1.4 **billion** years using Lead-Lead based isochron method.[19] A discrepancy of 1.2 billion years (or 765%)!

- Basalt from the Grand Canyon was dated as 516 million years by Geochron Laboratories, Cambridge, Massachusetts using an isochron method, 1.1 billion years by University of Colorado isotope laboratory, using a ribidium-strontium based isochron method, and 1.59 billion years by the same laboratory using a Samarium-Neodymium based isochron method.[20] A difference of 1.07 billion years!

These are just a few examples of the hugely disparate ages given to the *same rock* using isochron dating.

If all the current dating methods give such hugely divergent ages, how do scientists decide which age to publish? The answer may shock you. What tends to happen is that the age that is closest to their preconceptions, and is most convenient to their research, is accepted and the other ages are rejected. This is known as *"posterior reasoning"* and, sadly, it is widely practised.

For example, in 1994, scientists researching some Australopithecus Ramidus fossils at Aramis, Ethiopia, were trying to establish a date for them.[21] They sent samples of nearby basalt, closest to the fossil-bearing strata, for dating, and received the very inconvenient date of 23 million years. This did not accord with their evolutionary beliefs, which dictated that Australopithecus Ramidus did not exist that long ago. So they selected 26 basalt samples further removed from the fossils. Seventeen of these samples gave ages around 4.4 million years, and were accepted by the scientists. The other 9 samples gave ages far too old for their theory, so were rejected.[22]

This case is not at all unusual. When faced with an array of calculated ages, some scientists seem to select the age that is most convenient to their theory and reject the rest! Further examples illustrate the point:

- The primate skull known as KNM-ER 1470 (thought to belong to the Homo Rudolfensis family) was discovered in 1970, and various dates were assigned to it over a period of time, until an acceptable one was finally found.[23] Initially, the surrounding igneous rocks were dated as 230 million years. This was totally unacceptable to scientists, because humans did not exist that long ago. Over the years a series of descending ages were assigned to it until rocks were found that dated to only 2.9 million years. Although it was still on the high side of acceptability, it was agreed upon as the best date they could get. Scientists continued to be uneasy with such an old date until a recent study of pigs in Africa convinced anthropologists that the skull had to be much younger. More rock dating ensued, with many older ages being rejected, until some samples were discovered that gave a radiometric age of 1.9 million years. This is the best result that was obtained, and anthropologists have settled on this age.[24]

- Rocks from a volcanic deposit in Kenya, known as KBS Tuff, were originally dated at 230 million years old. However, this did not fit with the surrounding fossil strata which, according to the evolutionary timescale attributed to that strata, was much younger. Accordingly, the researchers continued to send new rock samples from other locations for dating until a suitable date of 2.6 million years was obtained![25]

If you think this seems like "dating lotto", you are correct. Such is the radiometric dating game. And it happens constantly when ages are assigned to fossils and rocks. In many cases, a scientist who is hoping for a particular age of a rock sample will ask the dating laboratory to only use the dating method that is likely to give a result closest to that age.

Similarly, the submission forms that radioisotope dating laboratories ask scientists to complete when submitting rock samples for testing, commonly ask how old the scientists are expecting or hoping the rocks to be.[26] One has to ask why dating laboratories feel the need to ask this question. If radiometric methods are completely reliable, and based upon objective, unquestionable science, such a question would be unnecessary. Obviously, the inclusion of this kind of question in a submission form is an admission of the widely variant ages that the different dating methods produce, and is a means by which dating laboratories can filter out anomalous dates that do not match the written brief.

**Huge Discrepancies for Rocks of Known Age**

We are not completely blind when it comes to dating rocks. Although the vast majority of igneous and volcanic rocks on earth were formed in the distant past, there are some rocks that have been formed in living memory. In the case of volcanic rocks, there have been numerous volcanic eruptions and lava flows in modern history that have resulted in newly formed rock when the lava has cooled. These are rocks whose ages we know for certain.

If our dating methods are a truly accurate and reliable means of determining the age of rocks, they should give accurate results for rocks whose ages are known. But this is not what happens at all.

What is both fascinating and deeply disturbing is what happens when samples of newly-formed rocks have been sent "blind" to dating laboratories around the world (meaning without informing the

# Rocks and the Age of the Earth

laboratories of the source and known ages of the rock samples). Without fail, any time this kind of blind testing of newly formed rocks has been conducted, the ages that have been calculated have been in the hundreds of thousands to hundreds of millions of years! Here are a few examples:

- Mt Ngauruhoe, New Zealand, has erupted several times over the last century; 1949, 1954 and 1975. On each occasion, lava flowed from the mountain and eventually cooled to become rock. In the late 1990s, rock samples from these lava flows were sent for radiometric dating to the PRISE Laboratory at the Australian National University, Canberra. The ribidium-strontium isochron gave an age of 133 million years. The samarium-neodymium isochron gave an age of 197 million years and the lead-lead isochron gave an age of 3.9 billion years.[27] Yet the rocks were only a few decades old!

- The Mt. Hulalai, Hawaii, erupted in 1801 and the lava flowed into the ocean where it cooled into rock. Some years ago, samples of this rock were blind tested, and the calculated age, using different methods, ranged from 1.9 billion years to 3.3 billion years.[28] Yet the rocks were only 200 years old!

- In 1973, a submarine volcanic eruption occurred off the coast of Japan. Over a period of 48 hours, a brand-new island, Nishinoshima Island, rose from the seabed. In the 1980s, blind testing was conducted on the rock samples from the newly formed island by several laboratories. The calculated dates ranged from 15 million years to 100 million years.[29] Yet the rocks were less than 20 years old!

- Mt. Krakatoa, Indonesia, erupted in 1883. In 1968, rock samples that were formed from the cooled lava flow were dated, and the calculated ages ranged from 12 million years to 21 million years. These results were published in "Science" magazine, in the October 1968 issue.[30] Yet the rocks were only 85 years old!

- In his article, *"A Critical Examination of Radioactive Dating of Rocks"*[31], Dr. S.P. Clementson cites several other instances where samples from rocks formed from recent volcanic eruptions in the Azores, Tristan da Cunha and Vesuvius were dated with ages ranging from 100 million years to 10.5 billion years. Yet these rocks were only centuries old!

These discrepancies represent errors in the magnitude of hundreds of millions of percent. And these are not a few isolated cases. Every time rocks of known age are dated using radiometric dating methods, these kinds of hugely exaggerated dates are given. **Every single time!**

How do evolutionists respond to these enormous dating errors of rocks with known ages? They argue that excess daughter isotopes were added to the rock as the lava flowed through the earth's upper mantle. Yet, as Dr. Eugenie C. Scott states;

> *"This is consistent with a young world, as the [daughter isotope] has had too little time to escape."*[32]

Furthermore, the argument by evolutionists that newly formed rock has been contaminated by daughter isotopes from mantle rock, effectively makes all radiometric dating meaningless.

As Dr. John F. Ashton states;

> *"If rocks known to be less than 100 years old, date as being hundreds of millions or billions of years old, how can we really know the age of **any** rocks from radiometric dating methods?"*[33]

How indeed! The validity of any scientific theory or methodology can only be maintained if it is consistently corroborated by testing against verifiable evidence in the observable world. In the case of radiometric dating, every test against verifiable data has shown it to be completely and laughably inaccurate.

If the testing of a theory or methodology consistently shows it to be in error, it must either be revised or discarded. It is grossly unscientific to hold on to a theory when every reality check shows it to be invalid. And yet, that is precisely what evolutionists are doing in regard to radiometric dating methods. They cling stubbornly to radiometric dating. Why? Because, without it, they cannot substantiate an old age for the earth, which is an essential foundation for the theory of evolution.

**HUGE DISCREPANCIES WITH CARBON 14 DATING**

The problems with radiometric dating become even more apparent when the apparent ages of rocks are compared with results from carbon 14 dating. Carbon 14 is an unstable radioactive isotope of carbon, found only in carbon-based life forms and in carbon-based objects such as coal and diamonds. Carbon 14 has a half-life of 5,730 years. In other words, after 5,730 years only half of the carbon 14 is left in any given sample. After two half-lives, one quarter is left. After three half-lives, one eighth is left. After ten half-lives, less than a thousandth is left. By these calculations, every atom of carbon 14 in a will have decayed after a few hundred thousand years. Most scientists believe the upper limit for carbon 14 to persist is only 100,000 years, and Dr. Eugenie C. Scott argues that anything older than 50,000 years should have no detectable C14 left.[34]

It stands to reason, therefore, that there should be no carbon 14 in fossils or diamonds found in rock strata that are dated as millions or even billions of years old. Yet this is not the case at all.

In the 1990s, Australian research geologist, Dr. Andrew Snelling, had some fossilized wood analysed for carbon 14. It had been found in rock strata dated between 40 million to 250 million years old. Clearly the fossilised wood should have contained no carbon 14. Yet it did! Using carbon 14 dating, the wood was dated at between 20,000 to 44,000 years old.[35]

This finding created a stir among scientists and caused a flurry of further investigation. Subsequently, further examples of fossils with carbon 14 have been found in strata that is dated as being millions of years old. Here are a few examples:

- In 2003, Drs. Baumgardner, Snelling, Humphreys and Austin, published findings of fossils in rocks dated up to 500 million years old by radioisotope dating methods, but which displayed C-14 ages of 20,000 to 50,000 years.[36]

- The same study revealed Precambrian (allegedly older than 545 million years) graphite, which is not of organic origin, which contained C-14.[37]

- Inspired by Snelling's initial findings in the 1990s, Dr. Paul Giem undertook a retrospective study of C-14 dating results from approximately 70 fossils dated from 1984 to 2001. Each of these had been assigned a C-14 age of 40,000 to 50,000 years, yet they were from rock strata dated from hundreds of thousands to millions of years old![38]

- No coal has ever been found that does **not** contain C-14, yet coal is supposed to be hundreds of thousands to millions of years old.[39] Scientists also consistently find C-14 in oil, natural gas limestone and marble from rock strata that is millions of years old, and this puzzling occurrence has been widely reported in conventional scientific journals.[40]

Once again, these are just a few samples of the flood of findings reporting measurable C-14 in fossils from strata dated in the millions and hundreds of millions of years old. In all of these cases, the occurrence of C-14 in fossils is *extremely* strong evidence refuting the supposed old age of the rock strata. It is simply *impossible* for the rock strata to be older than an absolute maximum of 100,000 years if C-14 is present.

How do evolutionists explain these finds of C-14? The initial response was to claim that contamination occurred. In other words, it was claimed that the researchers had somehow introduced C-14 to their samples during the testing process. As the number of puzzling C-14 finds multiplied, however, it became increasingly non-sensical to maintain this claim.

The final repudiation of the evolutionists' explanation was the discovery of C-14 in diamonds, in 2005.[41] Diamond is the hardest substance known, formed under immense pressure deep within the earth. As such, it is impervious to contamination. Diamond is supposedly extremely old. Secular geologists estimate that it was formed between 1 and 3 billion years ago. Given its extremely old age and its impervious nature, diamond should contain absolutely *no* C-14. Yet in 2005, geophysicist Dr. John Baumgardner, a member of the RATE research team (**R**adioactivity and the **A**ge of **T**he **E**arth), announced findings of measurable C-14 in diamonds he had studied.[42] The C-14 level in these diamonds was substantial - over ten times the detection limit of the dating laboratory's equipment.[43] Dr. Baumgardner then tested a further six alluvial diamonds from Namibia, and these had an even higher content of C-14.[44]

The presence of C-14 in diamonds is a stunning find for creationists and a major problem for evolutionists. It represents extremely strong evidence for a much younger earth and provides further strong data repudiating the ancient ages calculated by the now demonstrably unreliable radioisotope dating methods.

Furthermore, recent research by Dr. Eugenia C. Scott, a staunch evolutionist, proposes significant evidence that the current C-14 ages assigned to fossils may be considerably overstated.[45] Changes in the earth's atmosphere in the past, fluctuations in the earth's magnetic field, and the actions of a world-wide flood would have resulted in a much lower uptake of C-14 in the past and a much faster decay rate of C-14.

This means that fossils that seem to have a C-14 age today of 50,000 years could be as young as 6,000 years old. [46]

**OTHER EVIDENCE OF A MUCH YOUNGER EARTH**

Carbon 14 levels in fossils are not the only evidence for a much younger earth than radiometric dating indicates. Some of this evidence will be examined in greater detail in later chapters, but it includes:

- The recent discovery, by Dr. Mary Schweitzer, of dinosaur bones containing red blood cells, DNA and protein molecules, all of which can only last a few thousand years. Yet dinosaurs are supposed to have become extinct 65 million years ago![47]

- The discovery of DNA in a supposedly 120-million-year-old weevil preserved in amber.[48]

- Viable bacteria isolated from salt crystals in Permian strata which is supposedly 250 million years old.[49]

- Liquid blood extracted from a supposedly 300,000-year-old mammoth in Siberia.[50]

- Recent evidence of rapid deposition of multiple sedimentary rock layers; a process which was previously thought to have required millions of years. This will be examined in detail in the next chapter.

- The logical impossibility of the earth and the universe being billions of years old, given such things as erosion rates, the consumption rate of the sun's mass and other factors (which will be examined in Chapter 11, "*A Little Common Sense Please!*").

## CONCLUSION

If evolution is true, it absolutely requires billions of years in order for the supposed evolutionary processes to have even the remotest chance of occurring. However, the radiometric dating methodologies used to establish a long age for the earth are riddled with fatal flaws:

- They are based upon unprovable, uniformitarian assumptions which are increasingly being challenged by the latest research.

- They produce divergent and completely contradictory ages when the various radioisotope dating methods are compared with each other.

- They are wildly and laughably inaccurate in dating rocks of known age.

- They are completely contradicted by Carbon 14 residual in fossils.

- They are also contradicted by a range of other recent evidence, including the discovery of red blood cells, DNA and protein in fossils that are supposedly millions of years old.

Furthermore, the ridiculous practice of scientists selecting the most convenient date, from a list of hugely divergent dates presented to them by isotopic dating laboratories, cannot masquerade as objective science. Any reasonable, rational person must surely perceive this to be utter nonsense, rendering published dates completely meaningless.

Given the serious flaws in radiometric dating, and the overwhelming body of evidence contradicting it, one wonders how it can be perceived as a valid scientific methodology. One has to ask, *"How much more contradictory evidence is required before the scientific community abandons its faith in this wildly inaccurate, subjective dating methodology?"*

Of course, without it, the theory of evolution would fall apart. For this reason, evolutionists will fight, tooth and nail, to defend the current dating methodology, irrespective of the evidence levelled against it.

---

## ENDNOTES - Chapter 4

---

[1] https://sci.waikato.ac.nz/evolution/geological.shtml

[2] Dudley, H.C., *The Morality of Nuclear Planning*, Kronos Press in association with the centres of Interdisciplinary Studies, Glassboro State College, Glassboro, New Jersey, p. 61, 1976. / Dudley, H.C., *Chem. and Eng. News*, p. 2, 7 April 1975. / Read, J., *Chem. and Eng. News*, p. 5, 14 July 1975. / Emery, C.T., *Ann. Rev. Nuclear Science* **22**:165, 1972. / Anderson, J.L., *J. Phys. Chem.* **76**:3603, 1972 / Anderson, J.L. and Spangler, C.W., *J. Phys. Chem.* **77**:3114, 1973. /

[3] https://answersingenesis.org/geology/radiometric-dating/acceleration-of-radioactivity-shown-in-laboratory/

[4] https://www.purdue.edu/newsroom/research/2010/100830FischbachJenkinsDec.html

[5] https://www.forbes.com/sites/alexknapp/2011/05/03/radioactive-decay-rates-may-not-be-constant-after-all/#3f30758a147f

[6] John F. Ashton, "Evolution Impossible", Green Forest, AR, Masterbooks, 2013, p.143.

[7] F. Bosch, T. Fastermann, J. Friese, F. Heine, "Observation of Bound-state-b-Decay of Fully Ionized 187Re", Physical review Letters, Vol.77, no.26, 1996, pp.5190-5193

[8] W.K. Hensley, W.A. Bassett, J.R. Huizenga, "Pressure Dependence of the Radioactive Decay Constant of Beryllium-7", Science, Vol.181, No.4104, 1973, pp.1164-1165.

[9] L. Vardiman, A. A. Snelling, and E. F. Chaffin, eds., *Radioisotopes and the Age of the Earth: Results of a Young-Earth Creationist Research Initiative* (El Cajon, California: Institute for Creation Research; Chino Valley, Arizona: Creation Research Society, 2005); D. B. DeYoung, *Thousands . . . Not Billions* (Green Forest, Arkansas: Master Books, 2005).

[10] Don DeYoung, "Thousands, Not Billions", Master Books, Green Forest, Arkansas, 2005, pages 65–78

[11] https://answersingenesis.org/geology/radiometric-dating/radiometric-dating-problems-with-the-assumptions/

[12] John F. Ashton, "Evolution Impossible", Green Forest, AR, Masterbooks, 2013, p.140.

[13] G.R. Tilton, "Isotopic Ages of Zircon from Granites and Pegamites", Transactions, American Geophysical Union, Vol.38, 1957, p.364

[14] IBID

[15] IBID

[16] IBID

[17] George R. Tilton, "The Interpretation of Lead Age Discrepancies by Acid Washing Experiments", Transactions, American Geophysical Union, Vol. 37, 1956, p.226

[18] L.Vardiman, A.A. Snelling and E.F Chaffin, "Radioisotopes and the Age of the Earth", El Cajon, CA, Creation Research Society, 2005, pp.393f

[19] L.Vardiman, A.A. Snelling and E.F Chaffin, "Radioisotopes and the Age of the Earth", El Cajon, CA, Creation Research Society, 2005, pp.393f

[20] L.Vardiman, A.A. Snelling and E.F Chaffin, "Radioisotopes and the Age of the Earth", El Cajon, CA, Creation Research Society, 2005, pp.405-414

[21] G. WoldeGabriel, et.al. "Ecological and Temporal Placement of Early Pliocene Hominids at Aramis, Ethiopia", Nature 371, 1994, pp.330-333

[22] G. WoldeGabriel, et.al. "Ecological and Temporal Placement of Early Pliocene Hominids at Aramis, Ethiopia", Nature 371, 1994, pp.330-333

[23] Lubenow, M., The pigs took it all, Creation 17(3):36–38, 1995; creation.com/pigstook

[24] Lubenow, M., The pigs took it all, Creation 17(3):36–38, 1995; creation.com/pigstook

[25] F.J. Fitch and J.A. Miller, "Radioisotopic Age Determination of Lake Rudolf Artefact Site", Nature magazine, Vol.226, Issue 5242, April 1970, pp.226-228

[26] Eugenie C. Scott, "Evolution vs Creationism: An Introduction", University of California Press, Berkeley, 2nd edition, 2009, p.74.

[27] John F. Ashton, "Evolution Impossible", Green Forest, AR, Masterbooks, 2013, p.141.

[28] J.F. Evernden, D.E. Savage, G.H. Curtis, G.T. James, "Potassium-Argon Dates and the Cenozoic Mammalian Chronology of North America", American Journal of Science, Vol. 262, 1964, p. 145-198

[29] https://www.abc.net.au/news/2015-05-17/new-volcanic-island-off-japan-a-natural-lab-for-life/6476036

[30] "Science" magazine, 11th October, 1968.

[31] S.P. Clementson, "A Critical Examination of Radioactive Dating of Rocks", Creation Science Quarterly, Vol. 7, 1970, pp.137-141.

[32] Eugenie C. Scott, "Evolution vs Creationism: An Introduction", University of California Press, Berkeley, 2nd edition, 2009, p.74.

[33] John F. Ashton, op. cit. p.140

[34] Eugenie C. Scott, "Evolution vs Creationism: An Introduction", University of California Press, Berkeley, 2nd edition, 2009, p.67

[35] Vardiman, Snelling and Chaffin, editors, "Radioisotopes and the Age of the Earth", 2005, p.589.

[36] Baumgardner, J.R., Snelling, A.S., Humphreys, D.R. and Austin, S.A., Measurable 14C in fossilized organic materials: Confirming the young earth creation-flood model, Proc. 5th ICC, pp. 127–142, 2003.

[37] IBID

[38] Paul Giem, "Carbon 14 Content of Fossil Carbon", Origins, Vol.51, 2001, pp.6-30

[39] Eugenie C. Scott, "Evolution vs Creationism: An Introduction", University of California Press, Berkeley, 2nd edition, 2009, p.77

[40] https://answersingenesis.org/geology/carbon-14/carbon-14-in-fossils-and-diamonds/

[41] Vardiman, L., Snelling, A. and Chaffin, E., *Radioisotopes and the Age of the Earth*, Vol. II, ch. 8, Institute for Creation Research, California, USA, 2005.

[42] IBID

[43] https://creation.com/diamonds-a-creationists-best-friend

[44] IBID

[45] Eugenie C. Scott, "Evolution vs Creationism: An Introduction", University of California Press, Berkeley, 2nd edition, 2009, p.69

[46] IBID

[47] https://answersingenesis.org/fossils/dinosaur-tissue/

[48] https://www.ncbi.nlm.nih.gov/pubmed/8505978

[49] John F. Ashton, op. cit. p.131

[50] https://www.wired.co.uk/article/mammoth-blood

# Dirt, Rocks and Water

## Chapter 5

# Dirt, Rocks and Water

One of the most commonly cited "proofs" of evolution is the graduated fossil record throughout successive layers of sedimentary rock all over the world (known as the geological column). Scientists note that there is a marked progression from very small, simple fossils in the lowest sedimentary rock layers, to much larger, more complex fossils in the higher sedimentary layers. They conclude that this graduated record is the physical evidence of the gradual evolution of biological life, from the simple to the complex, over a large period of time. In reaching this conclusion, however, several major assumptions have been made:

1. Each individual layer, or strata, of sedimentary rock took long periods of time to form, with the whole geological column taking hundreds of millions of years to reach its current state.

2. The different strata represent different ancient epochs of the earth's geological and biological history.

3. Sedimentation rates have remained fairly constant over time (uniformitarianism).

4. There is no viable alternate explanation for the stratification of sedimentary rock and the gradation of fossils within sedimentary strata.

Not only are some of these assumptions speculative and unprovable, they are all increasingly questionable in the light of a growing body of contrary evidence.

## HISTORY

The idea that the observable strata of sedimentary rocks took millions of years to form took hold in the late 18th and early 19th centuries. The first step was assuming that because fossils in the higher strata were larger and more complex, they must represent a later period of biological development than the smaller fossils in lower strata. Thus, it was assumed that the vertical positions of the strata represent a timeline. Charles Lyell's 1991 seminal book, *"The Principles of Geology"*,[1] pioneered this concept, which was quickly embraced by scientists around the world. Lyell noted that each major strata of rock contained thousands, and even millions, of fine sedimentary layers, and he concluded that each of these layers must have formed very slowly and, consequently, that the whole geological column of strata represented hundreds of millions of years. Furthermore, Lyell assigned names to these different strata, designating a distinct geological time period - names which were embraced by the scientific world.

In the early stages of this theory, estimates of the ages of the different strata varied hugely, because they were based upon what can only be described as wild guesses of sedimentation rates. The more recent advent of radiometric dating, discussed in the previous chapter, allowed scientists to collaboratively agree upon clearer timescales, by dating volcanic rocks found at the same level as each sedimentary stratum.

The geological timescale that is currently proposed by evolutionists is indicated in the table below (also printed in the preceding chapter):

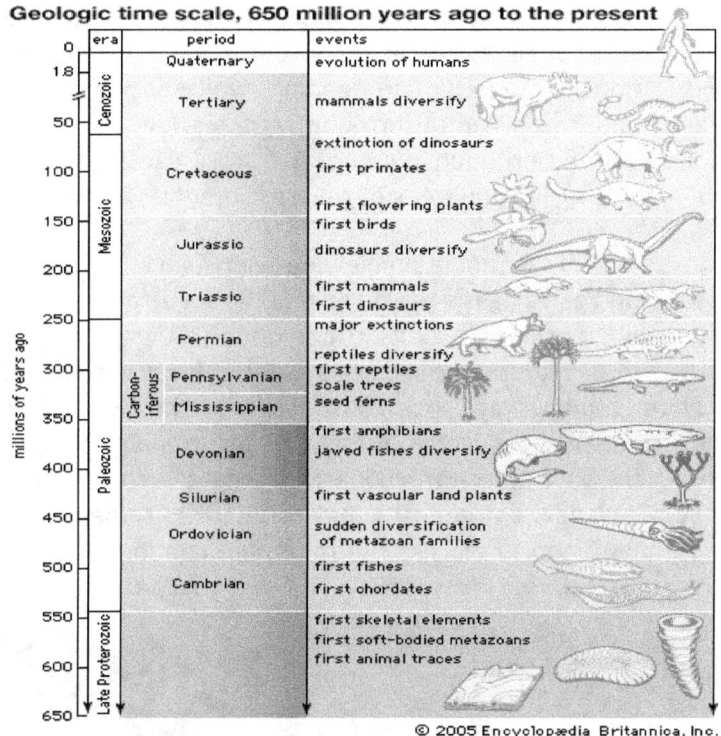

This generally accepted timescale of the geological column has become one of the foundations of evolutionary theory. The visual evidence of a gradation of fossils from small to large, simple to complex, in the geological column is regarded as unequivocal "proof" that evolution has occurred. But is this the only explanation for this kind of gradation? And do individual layers and stratum really take millions of years to form?

# Dirt, Rocks and Water

**PROOF OF RAPID SEDIMENTATION**

The 1980 eruption of Mount St. Helens, Washington State, USA, provided graphic evidence that multi-strata sedimentation can be laid down in a very short period of time. The eruption on May 18th, blew away the entire north slope of the mountain, sending a volcanic plume 24 km into the sky and depositing ash in 11 surrounding states. It was the largest and most violent volcanic eruption in American history, causing the deaths of 57 people and reducing hundreds of square kilometres to wasteland.[2]

As the north face of the mountain blew outwards, an initial superheated pyroclastic cloud of ash, gas, steam and mud flooded down the northern valley, initially travelling at 220 km per hour (350 mph) and quickly accelerating to 670 km per hour (1,080 mph)![3] This was soon joined by further water, mud, ash and rock as the glacier on the mountain melted. The landslide displaced huge amounts of water from Spirit Lake, to the north, in a giant wave 180 metre (600 ft) high that travelled further down the valley.[4] The sediment deposits that resulted from the eruption are 1.6 km (1 mile) deep in places. The mountain continued to erupt for several years afterwards in a succession of secondary eruptions.

**Before.**　　　　　　　　　　**After.**

Mount St. Helens, before and after the eruption. (source: vmf-214/Flickr)

As violently spectacular as all this was, the subsequent examination of the area revealed a stunning discovery. The flow of ash, water and sediment that flowed down the valley resulted in an 8 metre (25 ft) thick sedimentary deposit laid alongside the North Fork Toutle River. From photographic evidence, it is known that this whole deposit was laid down in only 3 hours, from 9 pm to midnight on Jun 12, 1980.[5] What is particularly surprising is that the newly laid sediment was deposited in perfectly defined layers called laminae, with layers of small particle sediment clearly separated from layers of larger particles.[6]

The formation of these kinds of finely layered sedimentary deposits had been conventionally believed to be an extremely slow process, taking hundreds of years for such laminae to form. But during the Mount St. Helens eruption, 8 metres (25 feet) of finely laminated layers was formed in just 3 hours! Furthermore, the eruption also proved that natural, cataclysmic events such as this can easily account for the graduated distribution of small to large particles and fossils within sedimentary layers and strata.

25 meters of newly layered sediments at Mount St. Helens © Morris, J. and Austin, S., *Footprints in the Ash*, Master Books, 2009

This rapidly formed, delineated sedimentation demonstrates a feasible, observable process whereby the graduated layers of fossils in other rock formations in the geological column could have been formed. This is strong evidence for the proposition that, rather than representing distinct geological ages, the delineated layers and strata in the earth's geological column may simply represent vertical sedimentary separation of animals of different sizes and weights as they were naturally distributed by a cataclysmic, rapid inundation (such as the biblical flood). Subsequent experiments in sedimentary deposition have demonstrated this kind of graduated layering as a naturally occurring rapid phenomena.[7]

Newly formed laminations at Mount St. Helens

The rapid formation of layered sediment at Mount St. Helens created a flurry of interest within scientific circles. A number of studies were subsequently conducted to ascertain whether rapid layering and gradation of sediment could be replicated under controlled laboratory conditions. One such study was conducted by Drs. P. Y. Julien, Y. Lan and G. Berthault, with their results published in scientific journals.[8] Their study provided verifiable proof that rapid stratification and sediment gradation can be achieved in hours, rather than hundreds, thousands or even millions of years.

This finding represents a serious challenge to one of the foundational beliefs of evolutionary theory. If the strata of the geological column were not formed in millions of years, the whole basis of evolutionary theory is in shambles.

## RAPID EROSION RATES

A second fundamental evolutionary belief that the eruption of Mount St. Helens overturned was the assumption that the erosion of canyons takes millions of years. Ongoing eruptions of the mountain caused flows of mud and water (from melted glaciers high up the mountain) to erode huge canyons at the base of the mountain. One of these canyons, dubbed "Little Grand Canyon", is 50 meters deep (160 ft) and 45 meters wide (150 ft), and was **carved in a single day**, on March 19th, 1982. It now has a stream flowing through it and, if you didn't know any better, you might assume that the canyon had been carved out over thousands of years.

'Little Grand Canyon' was carved by a catastrophic mudflow within a day. © Morris, J. and Austin, S., *Footprints in the Ash*, Master Books, 2009

Several other canyons were also created during Mount St. Helens' eruption phase, and provide further stunning evidence of extremely rapid erosion rates. Loowit Canyon is over 30 meters (100 ft) deep, and was carved partly through old, hard volcanic rock called andesite.[9] The canyon formed over a period of a few months in 1980. The largest of the canyons that was formed during the eruption phase is Step Canyon, over 200 meters (600 ft) deep. It was carved through old, hard volcanic rock (andesite) over a period of just a few months!

The Mount St. Helens eruption demonstrates that major geological reshaping of the earth's surface (geomorphology) can take place extremely rapidly. It must also be stated that the Mount St. Helen's eruption was a relatively small event when compared with known eruptions of the past. A paper on this topic by Creation Ministries International points out that the eruption of Vesuvius in 79 AD was three times larger than the Mount St. Helens eruption. The Krakatoa eruption (in what is now Indonesia, in 1883) was **18 times larger**, and the Tambora eruption (in 1815, also in Indonesia) was **80 times larger**.[10] The same paper indicates that the Deccan Trapps in India, a vast area of basalt rock from primordial (prehistoric) eruptions, is 5 million times the volume of the Mount St. Helens ejecta. Thus, the major volcanic activity known to have taken place in the distant past would have unleashed geomorphological forces hundreds, thousands and possibly even millions of times more powerful than those generated by the Mount St. Helens eruption. These kinds of cataclysmic events could easily account for the formation of much of the earth's geological contours, and upon examination of the evidence, this is a more convincing explanation.

Following the eruption of Mount St. Helens, Geologist, Dr. Steven A. Austin, spent several years studying its aftermath and comparing it to the morphology of the Grand Canyon. He concluded that there is now overwhelming evidence that the Grand Canyon was formed rapidly as a result of forces and processes of a similar but immensely larger scale to those at Mount St. Helens. He published his findings in 1994, in the classic work, *"Grand Canyon: Monument to Catastrophe"*.[11]

The commonly held belief that the earth's canyons and valleys were formed by slow processes over thousands and millions of years is based upon assumptions regarding the speed of erosion that are clearly challenged by this latest evidence. The geomorphology of the earth's surface can easily be explained as the product of a much younger earth, subjected to a period or periods of tremendous geological violence. A biblical reference to one such violent geological event is found in Genesis 7:11, which describes how *"the springs of the great deep burst forth,"* an event which would have presumably involved significant movement of the earth's tectonic plates.

**FOSSILISATION OF LARGE ORGANISMS**

A major inconsistency in the evolutionists' argument becomes evident when one considers the fossilisation of large animals such as the dinosaurs. According to evolutionary theory, sedimentation rates in the past were extremely slow, taking hundreds and thousands of years to form each laminated layer of sedimentary rock, with major strata taking millions of years. Yet, if this is true, how did dinosaurs and other vertebrates manage to be completely covered in sediment and fossilised? Surely if sedimentation was so ponderously slow, large animals would have completely decomposed and their bones would have eroded long before they could be completely buried and fossilised. The only logical conclusion is that these large animals were buried and fossilised very rapidly, in what must have amounted to a cataclysmic inundation event.

This is confirmed by the fact that we do not see large animals being fossilised today. If sedimentation and fossilisation in the past are supposed to have occurred gradually through processes that were uniformly equivalent to the same sedimentation rates that we experience today, why don't we witness fossilisation continuing to take place? We have now had hundreds of years of scientific observation and recording, yet we do not see masses of fossils being formed today. Crabs and starfish and seaweed are not being fossilised on our beaches. Muddy footprints are not being fossilised. Rabbits and beavers and kangaroos and bears are not being fossilised. Even recent localised flooding in the 20th and 21st century has not resulted in wildlife being buried and fossilised. Surely this is evidence that the mass fossilisation and sedimentation of the past required much more cataclysmic conditions than the gradual deposition rates that the current uniformitarian theory proposes.

Evidence of rapid burial is ubiquitous around the globe, including some extraordinary individual cases. For example, in 2011, palaeontologists in China discovered the fossil of an ichthyosaurus which perished in the process of giving birth to a baby.[12] This is strong evidence of rapid burial.

Ichthyosaurus giving birth.

Similarly, numerous fossils have been discovered of animals that have perished with half-swallowed prey in their mouths. These include a fish with a half-swallowed smaller fish in its mouth[13] and an ancient armoured fish which, in the words of an article on "Live Science" website, "was fossilized in the act of attacking and drowning a pterosaur in a toxic Jurassic lake."[14]

Fossilised fish eating a smaller fish.

Additionally, dinosaur and pterosaur skeletons are often found in what is known as the "opisthotonic posture" or "death throes posture", with neck arched and head thrown back.[15] In fact, skeletons found in this position are so common that a recent article in "New Scientist" made the comment that there are *"too many, in fact, to be a coincidence."*[16] Dr.

Marshall Faux points out that the opisthotonic pose results from severe malfunction of the central nervous system due to asphyxiation.[17] In other words, these animals died due to sudden suffocation.

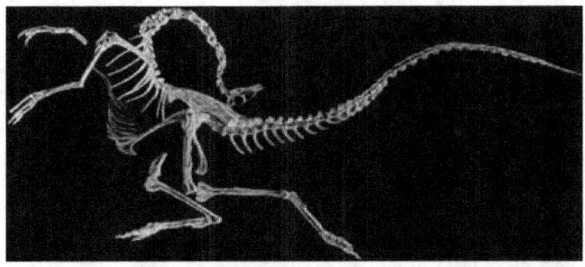

A struthiomimus in the classic opisthotonic death pose.

These and other instances are indicative of instantaneous death as a result of some major, cataclysmic event. A global flood with massive inundation of sediment would account for such rapid death and burial.

Further evidence of rapid burial abounds. There are mass fossil graveyards all over the earth which contain huge numbers of dinosaurs and other fossilised animals, apparently swept up together and buried rapidly under what appears to have been catastrophic conditions. Dr. Andrew Snelling cites a number of these massive fossil graveyards, including the Morrison Formation and associated deposits that stretch from New Mexico to Canada and cover an area of 1.5 million square kilometres.[18] This formation contains hundreds of thousands of well-preserved, at times pristine, fossils of dinosaurs such as stegosaurs and giant sauropods that grew up to 30 meters in length. It is important to stress that these were large land-dwelling animals that were buried in "*a massive conglomerate sandstone formation deposited by rapidly moving water.*"[19]

Dr. John F. Ashton describes a similar mass graveyard, the Soom Shale formation, in the Cedarberg Mountains of South Africa, which contain:

> *"thousands of exceptionally well-preserved fossils of brachiopods, nautiloids, anthropods and conodonts in locations spread over hundreds of miles. These fossils are so well preserved that the sensory organs, walking appendages, fibrous muscular masses and even gill tracts ...are remarkably preserved."*[20]

If sedimentation rates, and the resulting formation of sedimentary rock and shale, is extremely slow, as evolutionists claim, how did animals manage to be completely covered and fossilised, with soft tissue still intact, before they decomposed?

Similarly, Dr. Andrew Snelling describes the Cow Branch shale formations in Virginia and North Carolina, which contain huge numbers of fossilised animals with outlines of muscles and ligaments clearly obvious.[21] This can only have resulted if their burial and fossilisation occurred very rapidly before decomposition could occur.

The discovery of blood, haemoglobin, collagen, DNA and soft tissue such as cartilage in various dinosaur bones (which will be examined in detail in the next chapter) also clearly demonstrates the fact that fossilisation must have occurred rapidly, before this biological material decomposed. It is also extremely strong evidence that dinosaurs lived only a few thousand years ago, rather than millions of years in the past, as proposed by evolution.

## EVIDENCE OF CATASTROPHISM

Fossils cover the entire surface of the earth. There is no continent where they have not been found. There is no country, no significant land area, where fossils cannot be seen in abundance. They are on the tops of the highest mountains, at the bottom of the deepest valleys, in the middle of the driest desserts and on the floors of the deepest oceans. For example, the peak of Mount Everest contains marine limestone with fossils of small marine creatures embedded within the layers.[22] Fossils are only formed when a living or recently deceased organism is buried in water-born sediment, deposited as a result of inundation or flooding.

So how did fossils and sedimentary rock form on the tops of mountains and in all the earth's deserts? Because there is barely a mountain or desert in the world that has not been found to contain fossils.

Evolutionists explain this by postulating that, at different times, every land mass on earth was covered by ocean. Supposedly, all the earth's continents were once ocean beds that were thrust up to the surface by the action of tectonic plates. This theory would necessitate continents "popping up and down" at different times in the earth's distant past. However, there are several problems with this theory.

Firstly, Dr. Andrew Snelling points out;

> *"The continents are made up of lighter rocks that are less dense than the rocks on the ocean floor and rocks in the mantle beneath the continents. The continents, in fact, have an automatic tendency to rise, and thus "float" on the mantle rocks beneath, well above the ocean floor rocks. This explains why the continents today have such high elevations compared to the deep ocean floor, and why the ocean basins can hold so much water."*[23]

In other words, it is highly unlikely that today's continents were once ocean beds, because of their lighter consistency when compared to the denser structure of the existing ocean beds.

Secondly, there is the evidence of the fossils themselves. Although it is not widely publicised, fossils of both marine and land animals are commonly found mixed up together in the same rock layers all around the earth. If today's continents were once ocean beds which then rose up and became continents, there should be a marked separation of sea creatures and land animals in the fossil record, demonstrating these very different epochs. Yet this is often not the case!

For example, in 1987, a scientific expedition discovered fossils of whales and other marine animals high in the Andes Mountains.[24] Evolutionists explained this as evidence that the mountains had once been the bed of a shallow sea that had been pushed up slowly over time. Yet the expedition also found land animals mixed in with the marine fossils, apparently buried at the same time.[25] Similarly, there are fossils of land animals and marine creatures mixed together in the Himalayan Mountains and throughout Grand Canyon, all buried in the same layer at the same time.[26] The fossil evidence indicates that these land and marine animals all died *at the same time and were buried together*. This accords with the biblical description of a cataclysmic world-wide flood which would have caused the deaths and mixing together of land and sea creatures.

Similar mixed graveyards of marine and land animals are found in many other locations around the world. Some examples are:

- The Dinosaur National Monument in Utah, USA, where fossils of dinosaurs are mixed together with marine fossils in the same sedimentary layers.

- The Morrison Formation (previously mentioned). This is an area between New Mexico and Canada of approximately 1.5 million square kilometres. This massive graveyard contains fossils of fish, clams and other marine creatures all mixed together with fossils of dinosaurs, lizards, crocodiles, frogs and other land animals - all buried together at the same time!

- The largest fossil graveyard so far uncovered is the Karoo Formation in South Africa. Scientists estimate the number of fossils in this graveyard as being 800 million![27] And, once again, fossils of land and marine creatures lie side by side, all buried together and swept into a vast geological repository.

The fossil record in these and many other locations does not present distinct layers of marine fossils delineated from layers of land animals, supposedly separated by millions of years. Instead, they are mixed together, presenting stunning evidence of a singular cataclysmic event. A catastrophic world-wide flood is the best explanation for this, with vast graveyards being formed as the re-shaping of the earth caused the receding waters to sweep the carcases of dead creatures together.

Furthermore, the fact that sedimentary layers and strata span entire continents indicates flooding on a global scale rather than localised inundation. For example, the Tapeats sandstone layer spans most of North America, across North Africa and as far as Israel.[28] This continent-spanning, globe-encircling sedimentary layer appears to have been laid down in a single, global event - a flood of global proportions!

Further evidence of a global flood is found in the mineral and chemical composition of the sediment. For example, sandstone is comprised of fine particles of mineralised quartz (silicon dioxide) bound together with finely powdered cements of varying chemical compositions. When these minerals are chemically analysed, it is sometimes possible to identify where the grains of minerals and cements originated from. For example, in the case of the Tapeats sandstone layer in the Grand Canyon, the unique constituents of zircon, feldspar and other minerals and chemicals within the sandstone have enabled scientists to determine that these constituents had been eroded from the Appalachian Mountains and transported by flood waters 2900 km (1800 miles) across the continent! Once again, this is very convincing evidence that the sedimentary rock layers that are found all around the earth were not formed by small, localised flooding, or by ponderously slow ocean bed sedimentation, but by cataclysmic global flooding - the kind of flooding recorded for us in the Bible!

The evidence of continental sedimentary layers and huge mixed fossil graveyards testifies to a world-wide flood, and does not support the gradual deposition over millions of years proposed by evolutionists.

## POLYSTRATE FOSSILS

Polystrate fossils are fossils that span multiple layers or strata of sedimentary rock. The most common polystrate fossils are tree trunks, which are found standing upright (vertically) and spanning many layers of the geological column. Wikipedia concedes that fossilised polystrate tree trunks are found *"worldwide"*.[29] For example, the photo below is of a polystrate tree trunk in the Cumberland Basin, Nova Scotia. The fossilised tree spans multiple strata which, according to standard evolutionary theory, are supposed to take many thousands or millions of years to form.

Lycopsid, probably Sigillaria, with attached stigmarian roots. Specimen is from the Joggins Formation (Pennsylvanian), Cumberland Basin, Nova Scotia (Photograph by Michael C. Rygel).

This is only one of many thousands of examples worldwide. The problem for evolutionists lies in trying to explain how strata, which supposedly take thousands or millions of years to form, could slowly bury a tree trunk without the tree eroding away long before it could be completely buried. These polystrate fossilised trees defy the standard evolutionary view that strata took hundreds of thousands of years to form. Dr. Derek Ager, Professor of Geology at University College of Swansea, Wales, recognized this when he wrote of trees buried in coal seams in England:

> "If one estimates the total thickness of the British Coal Measures as about 1,000 m, laid down in about 10 million years, then, assuming a constant rate of sedimentation, it would have taken 100,000 years to bury a tree 10 m high, which is ridiculous!"[30]

Significantly, Derek Ager was *not* a creationist or a Christian, yet he could not deny the obvious implication that sedimentation and burial of fossils must have happened extremely rapidly on at least some occasions in the distant past.

So, what explanation do evolutionists offer? Wikipedia summarises the standard evolutionary response:

> "They [polystrate fossils] are formed by rare to infrequent brief episodes of rapid sedimentation, separated by long periods of either slow deposition, non-deposition, or a combination of both."[31]

This is certainly the consensus of evolutionary scientists.[32] This view of *occasional* rapid sedimentation is a convenient means of explaining evidence that contradicts the standard evolutionary view of slow sedimentation over hundreds of thousands of years. What evolutionists are effectively saying is, *"We believe that sedimentation occurred slowly over thousands of years, except when it obviously didn't."*

The problem with this view is threefold:

**Firstly**, there are not just a few exceptions where polystrate fossilised trees are found in the geological column, but literally thousands of them, found all around the world. Many of these trees appear to have been suddenly buried while still in the ground, as they display finely preserved fossilised root systems penetrating through many distinct layers of rock.[33] In fact, these polystrate tree trunks appear not just singly in the geological column; there are a number of sites, such as Nova Scotia, where there are whole forests of trees that have been inundated, buried and fossilised vertically in sedimentary strata.[34]

**Secondly**, the strata in which these polystrate fossils are buried is not evidently different *in any way* to strata elsewhere in the geological column. If, as evolutionists claim, strata with polystrate fossils was formed rapidly, and other strata was formed very slowly, there would be obvious differences in their mineral, chemical and geological composition, yet no such difference is observed.

**Thirdly**, trees are not the only form of polystrate fossils. A whale graveyard was recently discovered in Chile, with *thousands* of exquisitely preserved whale fossils spanning four strata of sedimentary rock. The whales were uncovered during excavation to widen the Pan-American Highway.[35] In explaining how the whale fossils are found across four separate strata, evolutionists theorise that there must have been four separate burial events, separated by millions of years. Yet, there is no evidence to support this at all! The whales are identical, showing no evolutionary change across the speculated millions of years, and they are all in exactly the same state of decay and preservation. Furthermore, several individual whale fossils lie at an angle of approximately 50 degrees, spanning several layers of sedimentary rock.[36] If these layers took thousands, or even merely hundreds of years to be deposited, the exposed portions of the whale carcasses would have completely decomposed and eroded long before they could have been buried!

**CONCLUSION**

Sedimentary rock has a story to tell; a very different story to the one proposed by evolutionists. It reveals:

- Rapid sedimentation and formation of laminated layers and strata.

- Rapid burial and fossilisation of large animals, some of which have beautifully preserved outlines of muscles, ligaments and soft tissue.

- Mass fossil graveyards with marine and land animals mixed together, all buried at the same time.

- Fossils of marine and land animals on every part of the earth's surface, including the world's highest mountains.

- Evidence of extremely rapid erosion and geomorphological shaping of the earth.

- Polystrate fossils, spanning many layers of sedimentary rock.

- Trans-continental and trans-global sedimentary layers.

- Sedimentary rock originating from sediment thousands of kilometres away.

The story that sedimentary rock tells is one of global, catastrophic flooding, rapid formation of strata and rapid fossilisation of animals. It tells the story of a young earth transformed by the biblical flood, and directly contradicts the long ages claimed by evolutionists. And without those long ages, evolution simply could not have occurred.

## ENDNOTES - Chapter 5

[1] Charles Lyell, "The Principles of Geography", 1991, University of Chicago Press.

[2] https://en.wikipedia.org/wiki/1980_eruption_of_Mount_St._Helens

[3] IBID

[4] IBID

[5] https://creation.com/lessons-from-mount-st-helens

[6] IBID

[7] https://creation.com/sedimentation-experiments-is-extrapolation-appropriate-a-reply

[8] Julien, P.Y., Lan, Y., and Berthault, G., "Experiments on Stratification of Heterogeneous Sand Mixtures", *J. Creation* **8**(1):37–50, 1994

[9] https://creation.com/lessons-from-mount-st-helens

[10] IBID

[11] Steve A. Austin, "Grand Canyon: Monument To Catastrophe", 1994, Institute For Creation research.

[12] https://www.newscientist.com/article/dn26975-stunning-fossils-mother-giving-birth/

[13] https://assets.answersingenesis.org/img/articles/am/v2/n2/fossil-fish.jpg

[14] https://www.livescience.com/18958-armored-fish-attacks-pterosaur.html

[15] https://creation.com/death-throes

[16] Lawton, G., The big sleep—Why are dinosaurs always found in the same position? There was only one way to find out … , *New Scientist* **196**(2635/2636): 62–63, 22/29 December 2007.

[17] Marshal Faux, quoted in https://creation.com/death-throes

[18] A.A. Snelling, "Earth's Catastrophic Past", Dallas, TX, Institute for Creation Research, 2009, Vol.2, p.487-577.

[19] John F. Ashton, "Evolution Impossible", Green Forest, AR, Masterbooks, 2013, p.74.

[20] IBID, p.75.

[21] A.A. Snelling, op. cit.

[22] https://www.thoughtco.com/geology-of-mount-everest-755308, March 2019,

[23] https://answersingenesis.org/fossils/fossil-record/high-dry-sea-creatures/

[24] https://www.nytimes.com/1987/03/12/us/whale-fossils-high-in-andes-show-how-mountains-rose-from-sea.html

[25] IBID

[26] https://answersingenesis.org/fossils/fossil-record/high-dry-sea-creatures/

[27] https://www.discoveryworld.us/geology/fossil-record-doesnt-support-evolution/

[28] https://answersingenesis.org/geology/rock-layers/sifting-through-layers-meaning/

[29] https://en.wikipedia.org/wiki/Polystrate_fossil, March 2019

[30] D.V. Ager, "The New Catastrophism: The Importance of the Rare Event in Geological History", Cambridge, 1993, Cambridge University Press. p.49.

[31] https://en.wikipedia.org/wiki/Polystrate_fossil, March 2019

[32] DiMichele, W.A., and H.J. Falcon-Lang, 2011, *Pennsylvanian 'fossil forests' in growth position (T0 assemblages): origin, taphonomic bias and palaeoecological insights.* Journal of the Geological Society, 168(2):585-605. / Gastaldo, R.A., I. Stevanovic-Walls, and W.N. Ware, 2004, *Erect forests are evidence for coseismic base-level changes in Pennsylvanian cyclothems of the Black Warrior Basin, U.S.A* in Pashin, J.C., and Gastaldo, R.A., eds., Sequence Stratigraphy, Paleoclimate, and Tectonics of Coal-Bearing Strata. American Association of Petroleum Geologists Studies in Geology.

51:219–238. / Archer, A.W., Elrick, S., Nelson, W.J. and DiMichele, W.A., 2016. *Cataclysmic burial of Pennsylvanian Period coal swamps in the Illinois Basin: Hypertidal sedimentation during Gondwanan glacial melt-water pulses*. In Contributions to Modern and Ancient Tidal Sedimentology: Proceedings of the Tidalites 2012 Conference: International Association of Sedimentologists.Special Publication (Vol. 47, pp. 217-231).

[33] https://rationalwiki.org/wiki/Polystrate_fossils, March 2019

[34] http://www.talkorigins.org/faqs/polystrate/trees.html

[35] http://bioweb.ie/chiles-whale-graveyard/

[36] https://rationalwiki.org/wiki/Polystrate_fossils, March 2019

Chapter 6

# Dinosaurs and Dragons

**THE COMMON MYTH**

The most common questions I am asked when presenting seminars on the creation / evolution debate, centre around dinosaurs. Young people, in particular, will raise their hands and ask, *"What about the dinosaurs?"*, to which I will respond, *"What about them?"*. The questioner will then ask something like, *"Do you believe they existed?"*, and I will reply, *"Of course they existed! There are hundreds of thousands of dinosaur skeletons and fossils all over the earth!"* This will almost invariably result in a degree of puzzlement in the questioner, who will then express something along the lines of; *"But dinosaurs are part of evolution, and you're saying evolution didn't happen!"*

# Dinosaurs and Dragons

As we start this topic, an important distinction needs to be made: dinosaurs are not part of the evolutionary story, they are part of the *earth's* story! They are part of the history of our planet! Dinosaurs are a small sub-set of extinct animals, among a plethora of millions of extinct species that once roamed the earth and subsequently died out for a variety of reasons. Where the theory of evolution and the creation account diverge is in the timing and manner of their extinction.

Unfortunately, it is the evolutionary narrative that dominates popular opinion. The picture is presented of dinosaurs roaming the earth millions of years before mammals arose, and many millions of years before humans evolved. They are depicted as being the predominant evolutionary lifeform at the time, reigning supreme in an exotic world of erupting volcanoes and strange reptiles. This view has been reflected in a veritable flood of children's animations and Hollywood movies. Published tables of the geological column and its associated ages reinforce this concept of the "age of the dinosaurs":

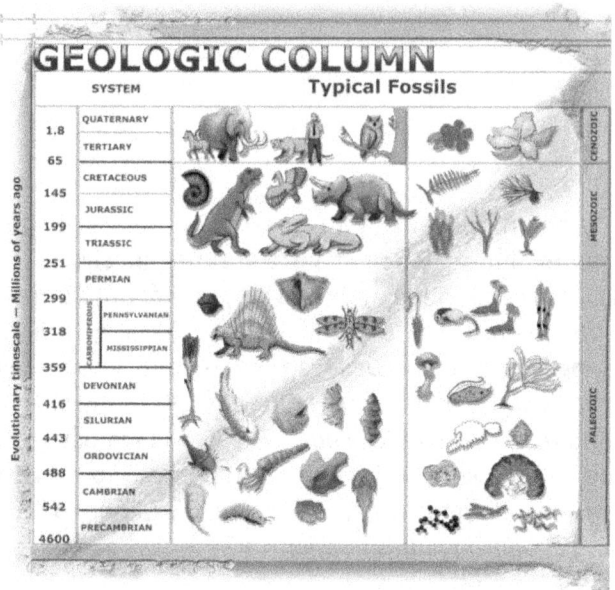

According to the accepted evolutionary timescale, dinosaurs died out 65 million years ago. It is believed that mammals evolved after the dinosaurs became extinct, and early pre-humans only appeared on the scene about 1.5 million years ago. This nice, neat evolutionary story, however, is quickly disintegrating in the wake of rapidly accumulating scientific evidence in recent decades.

**CO-EXISTENCE WITH MAMMALS**

Contrary to the attractively designed evolutionary geological tables, there is now convincing evidence that dinosaurs *did* exist side by side with mammals; they were not separated by millions of years as we have been led to believe. While evolutionists admit there was some cross-over (suggesting that early mammals began to emerge as the dinosaurs were dying out), the fossil record contradicts this gradual transition, giving evidence of a diversity of fully-developed mammals and other modern vertebrates living concurrently with the dinosaurs. Here are just a few examples:

- Duck fossils have been found in the Cretaceous strata along with dinosaur fossils.[1]

- In 2014, scientists from the American Museum of Natural History and the Chinese Academy of Sciences published data showing that three distinct species of squirrels existed in the so-called late Triassic period, along with the dinosaurs.[2]

- In 2008, Australian scientists published data confirming the existence of modern platypuses during the so-called age of the dinosaur.[3]

- The 24th February 2006 edition of *"Science"* journal, published findings of fully developed beavers living alongside dinosaurs.[4]

- In 2005, Chinese scientists discovered the fossil of a dog-sized mammal with the fossilised remains of a small dinosaur in its stomach.[5] This is proof that mammals hunted and ate dinosaurs over 130 million years ago (supposedly!) - long before such mammals are supposed to have evolved. Co-author of the published data in *"Nature"* journal, paleontologist Meng Jin commented, *"This new evidence gives us a drastically new picture"*.[6]

Dr. Donald Burge, palaeontologist with College of Eastern Utah Prehistoric Museum, states,

> *"We find mammals in almost all of our [dinosaur dig] sites. These were not noticed years ago .... We have about 20,000 pounds of bentonite clay that has mammal fossils that we are trying to give away to some researcher. It's not that they are not important, it's just that you only live once, and I specialised in something other than mammals. I specialise in reptiles and dinosaurs."*[7]

In an interview published in Creation magazine, Dr. Carl Werner stated that over 432 mammal species have now been identified in strata which was once supposed to be the domain of the dinosaur![8]

There is little doubt now, that mammals and dinosaurs did not exist in separate eras, but lived and competed concurrently, often dying alongside each other. The separation of the age of the dinosaurs from the age of the mammals by millions of years is a false evolutionary construct that is not supported by the fossil evidence.

**BIOLOGICAL EVIDENCE FOR THE RECENT EXISTENCE OF DINOSAURS**

In 1997, a team led by Dr. Mary Schweitzer, from Montana State University, extracted red blood cells with haemoglobin from the bones of a Tyrannosaurus Rex.[9] These samples were taken from deep inside a leg bone which had not completely fossilised, and is a strong indication of a very young age, rather than millions of years. The discovery of haemoglobin in dinosaur bones rocked the scientific world, because haemoglobin decomposes very rapidly (in comparative terms), and cannot last more than a few thousand years. In living organisms, haemoglobin has a lifespan of about 120 days before it becomes unviable and is replaced.[10] Once an animal dies, the red blood cells and haemoglobin molecules within the organism deteriorate at a steady, observable rate. This is because the protein molecules which comprise these elements are relatively unstable themselves, and deteriorate rapidly.[11] The process is further accelerated by microbes that quickly consume organic material.[12] For this reason it is an established scientific fact that red blood cells and haemoglobin molecules can survive for only a matter of hundreds of years, or a few thousand as an absolute maximum. This is why Dr. Schweitzer's stunning find was immediately so controversial, because it provided solid evidence that dinosaurs could not possibly be millions of years old.

In response to this, evolutionists claimed that Dr. Schweitzer's sample had been contaminated, and that it was impossible for dinosaur bones to contain intact biological elements. Dr. Schweitzer responded by conducting further research, excavating more dinosaur bones and examining their contents. Her subsequent studies were conducted under the most stringent, sterile regimes, and were scrutinised by a much larger group of independent scientists. In 2005, Dr. Schweitzer released even more extraordinary results from her studies. She had extracted still-flexible blood vessels from a Tyrannosaurus Rex bone and, from these structures, had been able to extract red blood cells![13] This was, once again, damning evidence, directly contradicting the evolutionary dinosaur narrative.

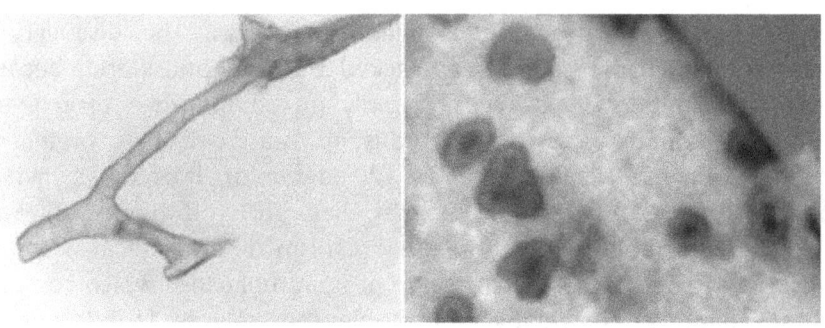

Blood vessels and red blood cells extracted from a T-Rex bone, 2005.[14]

Subsequent to her extraction of red blood cells, Dr. Schweitzer has also been able to extract protein molecules, collagen and blood enriched soft tissue from dinosaur bones.[15] After numerous attempts to discredit her work, her findings were eventually, and reluctantly, confirmed by the scientific community.[16] As a result, Dr. Schweitzer was awarded the Dr Elizabeth Nicholls Award for Excellence in Palaeontology, in 2018.[17]

Dr. Schweitzer's discoveries set in motion a flurry of further research. In the years that followed her initial discovery, a veritable flood of similar discoveries were made. These included DNA, soft tissue cartilage, protein molecules, red blood cells, and collagen - all extracted from dinosaur bones supposedly millions of years old.[18] For example, Prof. John Asara, of Harvard Medical School, was able to extract viable collagen from a T-Rex bone in 2007.[19] Other researchers subsequently found tendons, skin and ligaments still attached to some of the bones.[20] In some cases some of the soft tissue that was extracted was so fresh it was able to induce an antibody reaction to protein. In other words, it still retained some biological functionality![21]

How have evolutionists reacted to these extraordinary findings? Surely, they have rescinded their theory? Surely, they have renounced their evolutionary timescales and admitted that dinosaurs must have lived only thousands of years ago?

Not at all. In an extraordinary exhibition of rubbery science, evolutionists have simply attempted to adjust the facts to fit with their theory, by declaring that red blood cells, DNA and various soft tissues must be able to survive for millions of years after all! This is despite the fact that every piece of observable data in laboratories all around the world measurably reveals the rapid deterioration of these biological substances. It is an extraordinary piece of unscientific wishful thinking. True science requires that if the evidence does not substantiate a theory, the theory must either be modified to account for the evidence, or completely discarded if the evidence overwhelmingly contradicts the theory. In science, it is the theory that must be modified to fit the facts, not the other way around! Yet this recent development has seen evolutionists attempt to modify the facts to fit their theory.  This, perhaps more than any other development in recent decades, reveals the desperation of evolutionists to prop up their failing theory, and highlights the fact that evolution has increasingly become a faith-based belief system rather than an evidence-based scientific theory.

**HISTORICAL EVIDENCE FOR THE RECENT EXISTENCE OF DINOSAURS**

As well as stunning biological evidence for the recent existence of dinosaurs, there is also some intriguing historical evidence. For example, brass carvings on the tomb of Bishop Bell, in Carlisle Cathedral, England, clearly depict sauropods:

The above carvings, dated at 1496 AD, adorn the tomb of Bishop Bell, and show two sauropods engaged in battle. Another part of Bishop Bell's tomb features a brass carving of what is clearly a stegosaurus with a spiked tail. What is extraordinary about these carvings is that these dinosaurs were not known to modern scientists until the discovery of bones in the very late 19th century. How, then, could these carvings be made in the 1400s? There are only two possible explanations. Either these dinosaurs were still alive at the time (possibly nearing extinction) or, more likely, the carver had access to ancient documents which have since perished, which depicted these dinosaurs as witnessed by people from the past. Either way, the carvings at Carlisle Cathedral are strong evidence that humans had seen dinosaurs and recorded their likenesses long before modern scientists uncovered their skeletons.

In the ancient temples of Angkor, Cambodia, the pillars are adorned with stone carvings of animals. Among them is an undeniable carving of a stegosaurus:

Photo by Chris Maier, UnexplainedEarth.com

What is extraordinary about this carving is that it dates to about 1200 AD, long before modern palaeontologists had discovered the species. Once again, this is strong evidence that dinosaurs and humans had once coexisted.

These are not the only ancient depictions of dinosaurs. There are *many* other examples.

## Dinosaurs and Dragons

In 2017, pre-historic cave drawings of dinosaurs were discovered in a series of caves in Kuwait. The archaeologist who first examined them, Dr. Abdul Al-Shalafi, admits that his first assumption was that they were fraudulent, but subsequent radiocarbon analysis of the remnants of the paint, dated the drawings as 200,000 to 300,000 years old (in the evolutionary timescale).[22] Dr. Al-Shalafi comments:

> *"These prehistoric pictograms seem to illustrate dinosaurs being hunted by humans. This could mean that humans are actually to blame for their extermination. This is certainly a revolutionary new perspective and it will certainly be hard for many historians and archaeologists to accept."*[23]

Ancient examples of American Indian rock art, over 1,000 years old, have also been discovered in North America. A picture of a brontosaurus has been discovered at Bridges National Monument, Utah:[24]

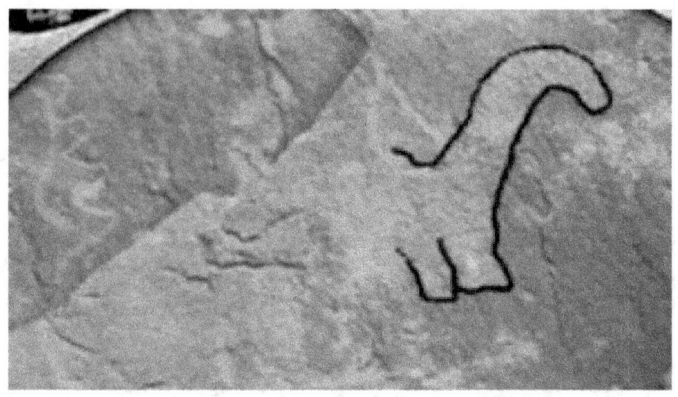

A T-Rex drawing has been discovered near Middle Mesa at the Wupatki National Park, Flagstaff Arizona:[25]

A stegosaurus painting adorns a rock wall at Lake Superior Provincial Park in Canada:[26]

Other examples of ancient American Indian rock paintings include several clear paintings of a brontosaurus:

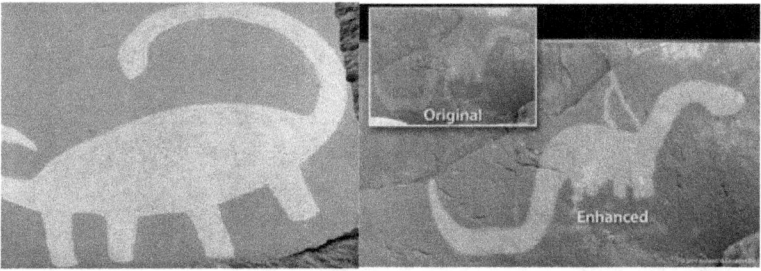

As well as rock art, there are literally hundreds of depictions of dinosaurs in pottery form. The examples below were discovered at the base of Toro Mountain, Mexico, in 1945, and are variously dated between 800 BC and 200 AD:[27]

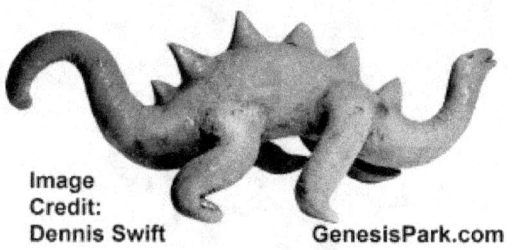

Image Credit: Dennis Swift          GenesisPark.com

## Dinosaurs and Dragons

There are many other locations around the world where excavations of ancient sites reveal pottery depictions of dinosaurs. The Sauropod, below, is part of a pottery vase, found in Ica, Peru, dated 100 BC:[28]

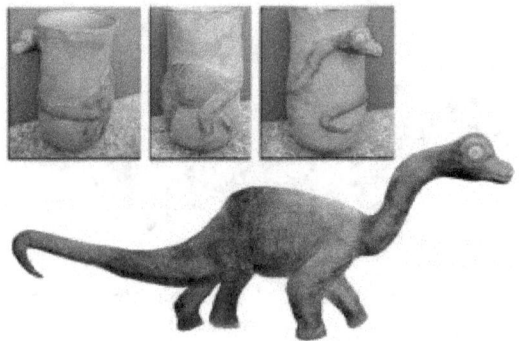

In the late 1800s, archaeologist Dr. Samuel Hubbard, Curator of Archaeology of the Oakland Museum, visited an area of the Grand Canyon known as the Havasupai Canyon. As an evolutionist, he was astonished to find several examples of ancient rock art depicting dinosaurs. He could not explain this, unless people had, indeed, once lived concurrently with dinosaurs. Writing of his discoveries, he stated:

> *"The fact that some prehistoric man made a pictograph of a dinosaur on the walls of this canyon upsets completely all of our theories regarding the antiquity of man. Facts are stubborn and immutable things. If theories do not square with the facts then the <u>theories</u> must change, the facts remain."*[29]

The only sensible explanation for ancient artistic depictions of dinosaurs all over the world is that the people who made them, or their ancestors, had seen these creatures themselves. Ancient dinosaur art is unassailable evidence that humans and dinosaurs coexisted, and that the evolutionary timescale is completely false.

The Anglo-Saxon epic, Beowulf, dated 500 AD, tells how Beowulf killed a reptilian monster named Grendel, along with its mother, but eventually lost his life to a flying reptile.[30] In this story, the description of the flying reptile that killed Beowulf is a precise depiction of a Pterosaur (a flying dinosaur). Grendel, the reptilian monster that Beowulf slayed, is described as standing on two large legs, having two small arms, a massive, powerful jaw with which it crushed its prey, and a hard skin that was impervious to sword blows. This is a perfect description of a Tyrannosaurus Rex, yet these were not discovered by modern scientists until the 19th century! While the anecdotal details of the story of Beowulf are clearly mythological, the precise descriptions of dinosaurs that it contains are further evidence of human encounters with such creatures in the real world.

## FIRE BREATHING DRAGONS?

While we are discussing ancient depictions of dinosaurs, we cannot ignore the overwhelming number of depictions of dragons. Almost every ancient culture that has been discovered displays drawings and pottery forms of these so-called mythical creatures. Artistic impressions of dragons feature very prominently in almost every culture on every continent on earth, from every time-period in human history.

Chinese drawing, dated 200 BC.[31]

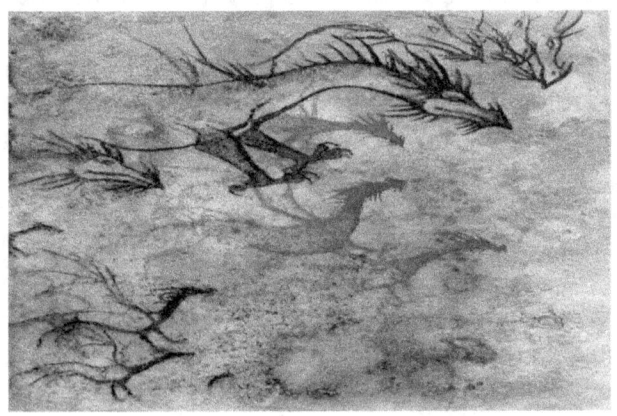

Chinese cave painting, dated 500-100 BC[32]

# Dinosaurs and Dragons

A fire breathing dragon relief in a French Chateau, dated early 1500s[33]

As well as artistic impressions, history is replete with stories of violent encounters with dragon-like creatures. The Mesopotamian epic of Gilgamesh, dated 2100 BC, describes Gilgamesh killing a huge dragon-like reptile in a cedar forest.[34] The early Britons describe ferocious, reptiles roaming the land.[35] King Morvidus, of Wales is reported to have been killed by a ferocious dragon-like reptile, in 336 BC.[36] Centuries later, King Peredur of Wales is reported to have slayed a dragon-like reptile at Llyn Llion, in 573 AD.[37] The article, *"Dinosaurs and Dragons: Stamping on the Legends"* at creation.com provides a rich source of other dragon tales from the ancient past.[38]

What are we to make of these stories, together with the multitudinous art depictions of dragons all over the world? Was there such a creature as a dragon? And, if such a creature existed, was it the fire-breathing reptile that is commonly described and frequently depicted? There are several points to consider.

# Dinosaurs and Dragons

**Firstly**, the term "dragon" came into usage about 1200 BC, derived from the Greek word, "drakon", meaning "huge serpent".[39] Prior to this, a variety of terms were used, such as monster and serpent.

**Secondly**, stories of encounters with dragon-like creatures appear in virtually all cultures around the globe, dating from 2,000 BC until the middle ages. What is remarkable is the fact that almost every ancient culture, despite being geographically isolated from each other, developed stories of fierce, reptilian monsters that were particularly aggressive towards people. Dragon-like reptiles are described and depicted in ancient literature and artefacts from a huge variety of cultures from antiquity: Mesopotamia (2,000 BC), Egypt (1,000 BC), Greece (1200 BC), Rome (200 BC), Russia and Ukraine (800 AD), Poland (1190 AD), Scandinavia (900 AD), Wales (11 AD), Britain (300 AD), India (1500 BC), China (2,000 BC), Japan (1400 BC), Peru (2,500 BC).[40] In fact, the Wikipedia page for "Dragon" cites about 150 stories of dragons or dragon-like creatures from ancient cultures all around the world.[41]

**Thirdly**, and following on from the previous point, the only reasonable explanation for such widely divergent and geographically isolated cultures developing extremely similar stories of dragons and ferocious reptilian monsters, is that such monsters actually existed in the past, and that humans regularly came into contact with them. The only remaining large, hostile reptile that exists today is the Australian saltwater crocodile. There is certainly nothing in contemporary existence that comes even remotely close to the ancient descriptions of large, reptilian predators. It may well be that terms such as "dragon" and "monster" were applied to a number of aggressive dinosaurs of the past, including the ferocious Tyrannosaurus Rex and the winged Pterosaur. This could account for variations in historical narratives and art depictions, with some "dragons" depicted as having wings, while others were wingless. The term "dinosaur" (meaning "large lizard") is a relatively modern addition our vocabulary, coined by Sir Richard Owen in 1841.[42] Prior to this, terms such as "dragon", "monster" and "serpent" seem to have prevailed as generic descriptors of large, hostile reptiles.

## Dinosaurs and Dragons

**Fourthly**, leaving aside those instances where the term "dragon" was applied to a variety of large, predatory dinosaurs, there does appear to be a consistency of description in a significant percentage of other cases. In these instances, dragons are uniformly described and depicted as large, elongated, serpentine creatures with four limbs, a spined or plated ridge along its back and tail, and a large bearded and horned or spiked head. Even more specifically, they are depicted as being covered in extremely tough, almost impenetrable scales. Descriptions and pictures of this creature do not seem to have an obvious correlation with any of the dinosaurs discovered to date. What is notable, is the remarkable correlation of descriptions and images of dragons in different ancient cultures, in completely separate regions of the globe.

Chinese dragon statue from the Han dynasty, 200 BC [43]

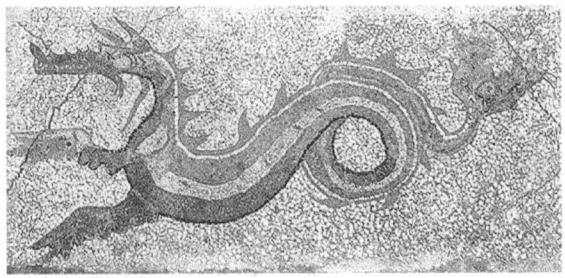

Ancient Greek Mosaic, from Caulonia, Italy[44]

# Dinosaurs and Dragons

**Fifthly**, there are references to dragons in the Bible. In the book of Revelation there are several references to dragons (Rev 12:1-17; 13:1-11; 16:13; 20:2). In Ezekiel 32, Pharaoh, King of Egypt, is referred to as a dragon or monster from the sea. The most remarkable biblical description of what appears to be a classic dragon is found in the book of Job, where "Leviathan" is described to Job by God himself:

> *"Can you pull in Leviathan with a fishhook or tie down its tongue with a rope? Can you put a cord through its nose or pierce its jaw with a hook? Will it keep begging you for mercy? Will it speak to you with gentle words? Will it make an agreement with you for you to take it as your slave for life? Can you make a pet of it like a bird or put it on a leash for the young women in your house? Will traders barter for it? Will they divide it up among the merchants? Can you fill its hide with harpoons or its head with fishing spears? If you lay a hand on it, you will remember the struggle and never do it again! Any hope of subduing it is false; the mere sight of it is overpowering. No one is fierce enough to rouse it. Who then is able to stand against me?*
>
> *Who has a claim against me that I must pay? Everything under heaven belongs to me. "I will not fail to speak of Leviathan's limbs, its strength and its graceful form. <u>Who can strip off its outer coat? Who can penetrate its double coat of armour?</u> Who dares open the doors of its mouth, ringed about with fearsome teeth? <u>Its back has rows of shields tightly sealed together; each is so close to the next that no air can pass between. They are joined fast to one another; they cling together and cannot be parted. Its snorting throws out flashes of light; its eyes are like the rays of dawn. Flames stream from its mouth; sparks of fire shoot out. Smoke pours from its nostrils as from a boiling pot over burning reeds. Its breath sets coals ablaze, and flames dart from its mouth.</u> Strength resides in its neck; dismay goes before it. The folds of its flesh are tightly joined; they are firm and immovable. Its chest is hard as rock, hard as a lower millstone. When it rises*

*up, the mighty are terrified; they retreat before its thrashing. <u>The sword that reaches it has no effect, nor does the spear or the dart or the javelin. Iron it treats like straw and bronze like rotten wood. Arrows do not make it flee; slingstones are like chaff to it</u>. A club seems to it but a piece of straw; it laughs at the rattling of the lance. Its undersides are jagged potsherds, leaving a trail in the mud like a threshing sledge. It makes the depths churn like a boiling caldron and stirs up the sea like a pot of ointment. It leaves a glistening wake behind it; one would think the deep had white hair. <u>Nothing on earth is its equal- a creature without fear</u>. It looks down on all that are haughty; it is king over all that are proud." (Job 41:1-34)*

This is a truly extraordinary description of a fire breathing dinosaur of some kind, with impenetrable scales and enormous strength. Could such a creature have existed? Or is this simply a mythological creature? The context of the book of Job gives every indication that this fire-breathing, serpentine creature actually existed at the time when Job lived. The description of Leviathan, in chapter 41, falls in the middle of a scathing discourse by God, as he rebukes Job for daring to question him. Over four consecutive chapters, God presents to Job a list of the wonders he has created, including astronomical and geological wonders as well as descriptions of a variety of living creatures. The latter include the ibis, rooster, lion, raven, mountain goat, deer, donkey, ox, ostrich, stork, horse, hawk and eagle. All of these creatures are listed by God as visual evidence of God's creative power. Job is invited to look upon these creatures and consider the greatness of the God who made them. It is in this context that God culminates his discourse with an entire chapter describing the fiercest of his creatures, the Leviathan. It would make no sense, whatsoever, if this final description was of an imaginary creature, when all those preceding it were creatures that Job could see and experience. In fact, throughout the description of this creature, God is at pains to point out how humans have tried to subdue it but have failed. Clearly, this was a creature that Job and his contemporaries were very familiar with.

This biblical account is an intriguing piece of evidence that a dragon-like creature once existed, and that it was more feared by humans than any other creature. This certainly accords with dragon stories from ancient cultures all around the world. These ancient accounts completely contradict any evolutionary notion of dinosaurs dying out millions of years before humans came on the scene!

But what are we to make of the notion of a creature that is capable of exhaling fire and smoke? Many people would suggest that this is complete nonsense. Yet we must take into account the fact that this description in Job 41 was enunciated by God himself. Surely God would not weaken his argument to Job by drifting into fanciful embellishment.

To date, no fire-breathing animals have been found, but it is not scientifically impossible for a creature to expel flames from its mouth, by storing a combination of volatile chemicals separately in its abdomen and expelling them in combination. The closest approximation of this process today is the bombardier beetle. The process is described on the "science.howstuffworks" website:

> *"For clues to a dragon's ability to breath fire, we turn to the real-life bombardier beetle. (Because who really wants to inspect a dragon's throat?) The bombardier beetle is a real-life expert at explosive spewing. The half-inch long beetle produces hydrogen peroxide and hydroquinones that are stored in separate reservoirs. When the beetle becomes threatened, it releases the hydrogen peroxide and hydroquinones into a special reaction chamber where secreted enzymes quickly break down the hydrogen peroxide and release free oxygen molecules that oxidate the hydroquinones. The result? A chemical reaction that makes enough heat to bring the entire mixture nearly to boiling."*[45]

Furthermore, biologists have proposed the possibility of animals producing self-igniting flames from their breath, via number of theoretical models. The website, scienceforstudents.org, examines several possible biological processes.[46] For example, methane and diphosphane gas can be produced by rotting vegetation in the gut, and diphosphane gas instantly ignites when brought into contact with oxygen. If a creature had the capacity to store these gasses separately, and in sufficient quantity, and could expel them through separate, simultaneous mechanisms, the result would be spectacular![47] There are at least four possible biological mechanisms whereby a creature could store and expel volatile gases that would burst into flame upon contact with oxygenated air.[48]

The Bible's description of Leviathan's fire-breathing ability is reinforced by many accounts from various ancient cultures. One of the oldest is the previously mentioned ancient Babylonian story of a hero named Gilgamesh who fought against a fire-breathing dragon. On several occasions the narrative of the story states, *"His mouth is fire, his breath is death."*[49]

Fire-breathing or not, however, there is certainly sufficient historical evidence for us to consider the plausibility of a fierce creature, commonly known as a dragon, to have once existed and to have terrorised people. The fact that we have yet to discover the remains of such a creature is not necessarily problematic, as scientists are regularly uncovering new, previously undiscovered species of dinosaurs. Given the fact that biologists estimate that up to 95% of all animals that once roamed the earth are now extinct, with some scientists, such as Dr. John F. Ashton, placing the estimate as high as 98-99%,[50] there remain millions of now-extinct species that we have yet to discover.

Furthermore, even if one is disinclined to believe in the existence of a specific creature known as a dragon, the multitudinous stories of human encounters with monsters who are given that name requires some explanation. Accounts with astonishing similarities from every corner of the globe and from almost every culture arguably constitutes overwhelming evidence of people encountering dragon-like dinosaurs of various kinds.

The evolutionary timescale that separates humans and dinosaurs by tens of millions of years is completely unsupported by the historical accounts and artistic records of many cultures.

**THE CAUSE OF EXTINCTION**

What killed the dinosaurs? Millions of children around the world are bitterly disappointed that they can't walk into their local pet shop and buy a pet dinosaur! So, what happened to them? Ralph may have broken the internet but, surely, he can't have killed all the dinosaurs!

Evolutionists are divided over the cause or causes of their mass extinction. Some of the most popular theories include:

- Toxic volcanic gases and dust
- Asteroid impact

- Food shortage

- Climate change

- Plants evolved poisons which killed the dinosaurs

- Poisonous insects evolved and stung the dinosaurs into extinction

- Cancer triggered by huge increases in solar radiation

While opinion is divided as to the exact cause of their demise, evolutionists are in general agreement regarding the timing; the last dinosaurs supposedly became extinct 65 million years ago. In this chapter we have already examined the overwhelming evidence that makes this evolutionary timescale a complete nonsense, but the question of the *cause* of their extinction still needs to be addressed.

Creationists often point to the biblical global flood as the cause of the dinosaur's extinction, but this is an oversimplification. The global flood certainly accounts for the many mass fossil graveyards that are found all over the earth, containing dead animals and organisms of every kind. It also makes sense of the fact that fossils cover the entire surface of our planet, from the highest peaks to the driest deserts. The Bible describes the catastrophic, global death that the flood caused:

> *"The waters rose and covered the mountains to a depth of more than fifteen cubits. Every living thing that moved on land perished—birds, livestock, wild animals, all the creatures that swarm over the earth, and all mankind. Everything on dry land that had the breath of life in its nostrils died. Every living thing on the face of the earth was wiped out; people and animals and the creatures that move along the ground and the birds were wiped from the earth. Only Noah was left, and those with him in the ark." (Gen 7:20-22)*

We have already noted the fact that a cataclysmic global flood and its aftermath, with accompanying geological reshaping of the earth's surface, is the best explanation for the rapid burial in sediment of large numbers of huge animals all over the earth. The largest "bone-bed" fossil graveyard in the world is located in Northern Montana, USA, where an estimated 10,000 duckbill dinosaurs have been swept together and suddenly buried. The massive forces involved in such inundation and burial continue to perplex evolutionists. For example, commenting on the evidence of violent sedimentary deposition in this Wyoming graveyard, Drs. J. R Horner and J. Gorman comment:

> *"How could any mud slide, no matter how catastrophic, have the force to take a two- or three-ton animal that had just died and smash it around so much that its femur—still embedded in the flesh of its thigh—split lengthwise?"*[51]

Similarly, Dr. Robert Bakker comments on another fossil graveyard in Eastern Wyoming:

> *"Anyone who cherishes notions that evolution is always slow and continuous will be shaken out of his beliefs by Breakfast Bench [Como Bluff] and the other geological markers of cataclysm."*[52]

But the biblical global flood does not account for every fossil in the geological column, nor does it explain the extinction of the dinosaurs as a whole. In regard to the fossil record in the geological column, animals and organisms were dying and being fossilised long before the biblical flood. While only something as catastrophic as a global flood can result in large animals, such as dinosaurs, being buried and fossilised, smaller organisms which died prior to the flood could easily have been fossilised in smaller, localised sedimentary events over a period of time.

More importantly, the global flood does not explain the complete extinction of the dinosaurs, because the Bible records that God saved pairs of all the animals on the ark, and that they subsequently repopulated the earth:

> "On that very day Noah and his sons, Shem, Ham and Japheth, together with his wife and the wives of his three sons, entered the ark. 14 They had with them every wild animal according to its kind, all livestock according to their kinds, every creature that moves along the ground according to its kind and every bird according to its kind, everything with wings. 15 Pairs of all creatures that have the breath of life in them came to Noah and entered the ark. 16 The animals going in were male and female of every living thing, as God had commanded Noah. Then the Lord shut him in." (Gen 7:13-16)

It can be assumed, therefore, that dinosaurs were among the animals that went aboard the ark and were saved for the preservation of their species. This accords with the previously mentioned historical accounts of human interactions with dinosaurs in the not-so-distant past. If this is true, two questions immediately come to mind: How did the dinosaurs (and, indeed, all the animals) fit on the ark? And, secondly, why are there no dinosaurs today?

**How Did the Dinosaurs Fit On The Ark?**

**Firstly**, Noah's Ark was spectacularly huge! The dimensions recorded in Genesis 6, give it a total volume of 42,475 cubic meters (1.5 million cubic feet).[53] It was as long as one and a half football fields and as tall as a four-story building. With its multiple decks, it had a total floor area of 16,640 square meters (173,400 square feet). Furthermore, the length to width ratio of 6:1 is reportedly the most stable ratio for large vessels, and would have made the ark extremely sea-worthy and almost unsinkable.[54]

**Secondly**, Noah didn't take two of *every single species* onto the ark. He was instructed to take two of every "*kind*" (Gen 6:19-20). With our modern understanding of genetics, this makes perfect sense. Noah did not need to have every breed of dog on the ark. A pair of wolves would have had all the necessary genetic information to diversify into the various dog breeds that we see today. Similarly, only one or two breeds of cat would be all that was necessary to diversify into the varieties of wild and domesticated cats that we see today. The same is true of all the animal species. The process of diversification through genetic specialisation is a phenomenon that is now well-understood, and one which we see in operation today with the successful development of new breeds of plants and animals. Obviously, Noah did not have the scientific knowledge to understand all this, but God did! This is why, in Genesis 7, it is God who brings all the necessary animals to Noah, for him to place in the ark.

**Thirdly**, Noah did not need to bring *adult sized* specimens into the ark. There is unequivocal fossil evidence that even the largest dinosaurs started life very small. The largest of the dinosaurs and other animals could easily have been represented on the ark by very young or juvenile specimens.

All of these factors make the account of the global flood and Noah's Ark completely feasible. They also provide us with a logical explanation of how dinosaurs could have survived on the earth after the flood, thus giving rise to the various written accounts in the historical record.

But it still does not explain why there are no dinosaurs today.

### Why Are There No Dinosaurs Today?

Clearly, dinosaurs faded into extinction sometime after the flood. If biblical historians are correct in dating the global flood at around 4,300 years ago, dinosaurs became extinct sometime during the subsequent years. Their extinction seems to have been a very gradual process, as evidenced by the written historical record. The historical accounts mentioned earlier in this chapter indicate that humans in the past encountered large dinosaurs, but these historical accounts became more infrequent as time went on. Historical records of human encounters with dinosaurs or "dragons" fade to a trickle by the time of Christ, and completely cease by the end of the middle ages. It appears that by the time of Christ, only a few, long-lived specimens survived, probably in isolated geographical pockets, and these eventually perished, either through insufficient numbers to sustain a viable population or through devastating interaction with humans (or a combination of both).

This gradual demise of the dinosaurs is further attested by the fossil evidence. A paper published on the Genesis Park website on the topic of the gradual decline of dinosaurs in the fossil records, quoting the research of Dr. Timothy Clarey,[55] states:

> *"But perhaps the biggest puzzle of the dinosaur extinction is the pattern of the fossils. While few dinosaurs seemed to have been buried above the so called "K/T boundary" that marked the end of the Mesozoic Era, the pattern doesn't fit a mass catastrophic extinction at just that point. One would expect to see (if perhaps a meteor strike had killed them all) an occasional fossil in the lower layers, a mass of fossils in the layer close to the extinction event, and then none thereafter. Instead we see many dinosaur fossils through the Mesozoic Era, trailing off to a very few in the top layer. In the famous Hell Creek Formation, for example, this "gradual demise" is highlighted by a 2-3 metre gap in the top of the Cretaceous where there are only a few fragments and no complete skeletons found at all."*[56]

So, what caused this gradual decline of the dinosaurs?

The extinction of the dinosaurs needs to be put into perspective. The extinction of various species is an ongoing occurrence in our world. Biologists estimate that somewhere between 8 and 270 species of insect, bird and mammal become extinct every 24 hours.[57] Yes, that is not a misprint! Every 24 hours! The reason for the huge range of this estimate (8 - 270) is that biologists have no clear idea how many species of biological life currently exist on our planet. Estimates range from 3 million to 100 million species currently in existence. The estimated yearly biodiversity loss through extinction is approximately 0.1%, giving an extinction rate of between 3,000 and 100,000 species each year.[58]

The fact that a few hundred, or even a few thousand, species of dinosaur no longer exist is "chicken feed" compared to the huge number of species that have faded into extinction over the millennia. As previously mentioned, Dr. John F. Ashton comments that biologists estimate that up to 99% of all species that once lived upon the earth are now extinct.[59] A significant factor in this would be the global flood which wiped out all life upon the earth except for those specimens taken on board the ark in order to ensure the survival and future genetic diversification of their "kind".

The aftermath of the global flood would have resulted in significant geological and environmental changes upon the earth, which would have further impacted the survival of various species. Some species would have flourished in the changed conditions, while others would have struggled and eventually perished. It appears that dinosaurs were in the latter category.

Dr. Larry Vardiman, states:

> *"If the Flood really happened as described in Genesis, then the earth's atmosphere, ocean, and crust would have been severely disrupted. Following the Flood, it would have taken thousands of years for these systems to return to a new equilibrium and even then, these systems would continue to exhibit slow changes."*[60]

Two large-scale field tests in 1993 and 1995 established the strong likelihood that a global flood may have initially resulted the cooling of the oceans and the advent of an Ice-Age. Significantly increased levels of iron in the oceans may then have led to a period of global warming which would have melted the ice-caps and ended the Ice-Age.[61] Interestingly,

God's question to Job, in Job 38:2,9 could be an oblique reference to a post-flood ice-age; *"From whose womb comes the ice?"*

In these kinds of radically altered environmental conditions, many species would have struggled to survive. Those that could adapt, survived, while those that could not, faded into extinction. The latter is the likely fate of the dinosaurs.

**DINOSAURS AND THE BIBLE**

When I conduct seminars on the topic of creation versus evolution, I am sometimes asked, *"Why aren't dinosaurs mentioned in the Bible?"*, to which I reply, *"They are!"* What needs to be understood, however, is that the term "dinosaur" (meaning *"terrible lizard"*), was not coined until 1841, by Sir Richard Owen. The Bible pre-dates this term by thousands of years, so we should not expect to find it written within the pages of scripture. What we do find in the Bible, however, are the ancient names and descriptions of specific dinosaurs, as well as the generic term "dragon" (previously discussed), which was sometimes used in the ancient world to refer to the general genre of large, predatory dinosaurs.

We have already examined the detailed description, in the book of Job, of the large, fierce creature known in the ancient world as "Leviathan". There are three further references to Leviathan in the Bible; Psalm 74:13-14, Psalm 104:26 and Isaiah 27:1. In these verses, Leviathan is described as *"the monster in the waters"* (Psa 74:13), who *"frolics in the sea"* (Psa 104:26) and *"the gliding serpent ... the coiling serpent ... the monster of the sea"* (Isa 27:1). Whether this was a dragon in the classic sense or not, there seems to be little doubt that this was some form of dinosaur that is now extinct.

Job chapter 40 also mentions another specific dinosaur, "Behemoth":

> *"Look at Behemoth, which I made along with you and which feeds on the grass like an ox. What strength it has in its loins, what*

*power in the muscles of its belly! Its tail sways like a cedar tree; the sinews of its thighs are close-knit. Its bones are tubes of bronze, its limbs like rods of iron. It is the greatest among God's creatures." (Job 40:15-19)*

This is obviously an enormous, herbivorous creature with a tail like a cedar tree; the biggest land animal that God created. While we cannot be certain which specific species the Behemoth was, several large herbivorous species have been suggested, including the diplodocus and brachiosaurus (now renamed "giraffatitan brancai").

Diplodocus     Brachiosaurus

While we cannot be certain which species the Behemoth was, there is little doubt that it was a dinosaur that is now extinct. Certainly, there is no creature alive today that can accurately be described as having a tail *"like a cedar tree"* (Job 40:17).

Strangely, however, the footnotes in some Bibles, in attempting to explain the reference to Behemoth in Job 40, state, *"possibly a hippopotamus or elephant"*! The Bible translators appear to have been thoroughly duped by evolutionary timescales to believe that dinosaurs could not possibly have cohabited the earth with humans. Therefore, their best explanation of Behemoth is that it may have been a hippopotamus or elephant!

REALLY? Have you ever seen a hippo's tiny tail? Or an elephant's? Do their tails remind you of a cedar tree? I don't think so!

## CONCLUSION

The term "dinosaur" is a generic term used to refer to a number of species of now extinct animals, among many millions of other extinct animals. The various fossil strata record their existence and their gradual demise, along with mass graveyards resulting from at least one episode of cataclysmic extermination. Although evolutionists claim that dinosaurs became extinct millions of years before humans existed, this claim is contradicted by a variety of evidence. Recent biological evidence, including the discovery of red blood cells, DNA, collagen, and soft tissue within dinosaur skeletons, makes the claim of millions of years an impossibility. Historical evidence, including ancient depictions and written accounts, clearly indicates the concurrent existence of humans and dinosaurs in the not-too-distant past. Furthermore, the Bible includes descriptions of dinosaurs, existing concurrently with humans.

Dinosaurs are part of the story of the earth; created by God, buried in the flood, preserved via the ark and eventually fading into extinction, quite possibly as a result of human intervention. Just don't believe the fanciful spin woven into their story by evolutionists!

## ENDNOTES - Chapter 6

[1] "Cretaceous Duck Ruffles Feathers", *BBC news*, www.bbc.co.uk, 20 January 2005

[2] https://www.sciencedaily.com/releases/2014/09/140910132522.htm

[3] "Swimming with Dinos", www.museumvictoria.com.au, 24 January 2008, accessed 1 October 2010.

[4] Early Aquatic Mammal, *Science* **311** (5764): 1068, 24 February 2006

[5] https://www.livescience.com/3794-dinosaur-fossil-mammal-stomach.html

[6] IBID

[7] Interview with Dr Donald Burge, curator of vertebrate paleontology, College of Eastern Utah Prehistoric Museum by Dr Carl Werner, 13 February 2001, in *Living Fossils— Evolution: The Grand Experiment, Vol. 2,* New Leaf Press, 2009, p. 173

[8] "Living Fossils: A Powerful Argument for Creation", *Creation* **33**(2):20–23, 2011.

[9] https://en.wikipedia.org/wiki/Mary_Higby_Schweitzer

[10] https://mytransfusion.com.au/reasons-transfusion/anaemia

[11] https://www.iflscience.com/plants-and-animals/scientists-may-have-found-red-blood-cells-75-million-year-old-dinosaur-fossil/

[12] https://www.iflscience.com/plants-and-animals/scientists-may-have-found-red-blood-cells-75-million-year-old-dinosaur-fossil/

[13] https://creation.com/sensational-dinosaur-blood-report

[14] IBID

[15] https://en.wikipedia.org/wiki/Mary_Higby_Schweitzer

[16] San Antonio, James D.; Schweitzer, Mary H.; Jensen, Shane T.; Kalluri, Raghu; Buckley, Michael; Orgel, Joseph P. R. O. (2011-06-08). Van Veen, Hendrik W., ed. "Dinosaur Peptides Suggest Mechanisms of Protein Survival". PLoS ONE. / Also https://en.wikipedia.org/wiki/Mary_Higby_Schweitzer

[17] https://en.wikipedia.org/wiki/Mary_Higby_Schweitzer

[18] John F. Ashton, "Evolution Impossible", op. cit. pp.78-83.

[19] https://www.reuters.com/article/us-dinosaur-idUSN1231500620070413

[20] https://www.bbc.com/news/science-environment-33067582

[21] IBIB, p.79.

[22] http://earth-chronicles.com/histori/in-kuwait-found-rock-paintings-of-people-and-dinosaurs.html

[23] https://worldnewsdailyreport.com/prehistoric-cave-art-depicting-humans-hunting-dinosaurs-discovered-in-kuwait/

[24] https://www.discoveryworld.us/dinosaur-world/indian-rock-art/

[25] IBID

[26] IBID

[27] https://www.genesispark.com/exhibits/evidence/historical/ancient/dinosaur/

[28] https://www.genesispark.com/exhibits/evidence/historical/ancient/dinosaur/

[29] Doheny, E. L., *Discoveries Relating to Prehistoric Man by the Doheny Scientific Expedition in the Hava Supai Canyon Northern Arizona*, 1924, p. 5.

[30] https://creation.com/dinosaurs-and-dragons-stamping-on-the-legends

[31] https://i.pinimg.com/236x/22/02/60/220260a6aaa6a0483e506c697e5e156b--chinese-dragon-dragon-art.jpg

[32] https://www.pinterest.com.au/pin/384283780680956940/?lp=true

[33] https://www.genesispark.com/exhibits/evidence/historical/ancient/dinosaur/

[34] https://en.wikipedia.org/wiki/Epic_of_Gilgamesh

[35] Cooper, B., 'Anglo-Saxon Dinosaurs described in early historical records', pamphlet No.280, Creation Science Movement, Portsmouth, UK, 1992.

[36] https://creation.com/dinosaurs-and-dragons-stamping-on-the-legends

[37] https://www.revolvy.com/page/Peredur-son-of-Efrawg

[38] https://creation.com/dinosaurs-and-dragons-stamping-on-the-legends

[39] https://en.wikipedia.org/wiki/Dragon

[40] https://en.wikipedia.org/wiki/Dragon

[41] https://en.wikipedia.org/wiki/Dragon

[42] https://creation.com/dinosaurs-and-dragons-stamping-on-the-legends

[43] https://www.genesispark.com/exhibits/evidence/historical/ancient/dinosaur/

[44] https://en.wikipedia.org/wiki/Dragon

[45] https://science.howstuffworks.com/science-vs-myth/strange-creatures/could-dragons-really-breathe-fire.htm

[46] https://www.sciencenewsforstudents.org/blog/technically-fiction/nature-shows-how-dragons-might-breathe-fire

[47] https://www.sciencenewsforstudents.org/blog/technically-fiction/nature-shows-how-dragons-might-breathe-fire

[48] https://www.sciencenewsforstudents.org/blog/technically-fiction/nature-shows-how-dragons-might-breathe-fire

[49] https://creation.com/behemoth-and-leviathan

[50] John F. Ashton, Evolution Impossible", Green Forest, AR, Masterbooks, 2013, p.73

[51] Horner, J.R. and Gorman, J., *Digging Dinosaurs*, Workman Publishing, New York, 1988.

[52] Bakker, R.T., *The Dinosaur Heresies—New Theories Unlocking the Mystery of the Dinosaurs and Their Extinctions*, Kensington Publishing Co., New York, pp. 425–444, 1986

[53] http://www.biblestudy.org/basicart/was-noah-ark-big-enough-to-hold-all-animals.html

[54] https://wol.jw.org/en/wol/d/r1/lp-e/102007008

[55] Clarey, Timothy, "The Hell Creek Formation: The Last Gasp of the Pre-Flood Dinosaurs," *CRSQ 51:4,* 2015, p. 294

[56] https://www.genesispark.com/exhibits/trivia/killed/

[57] http://wwf.panda.org/our_work/biodiversity/biodiversity/

[58] http://wwf.panda.org/our_work/biodiversity/biodiversity/

[59] John F. Ashton, "Evolution Impossible", op. cit. p.73.

[60] https://www.icr.org/article/global-warming-flood/

[61] https://www.icr.org/article/global-warming-flood/

# Irreducible Complexity

Chapter 7

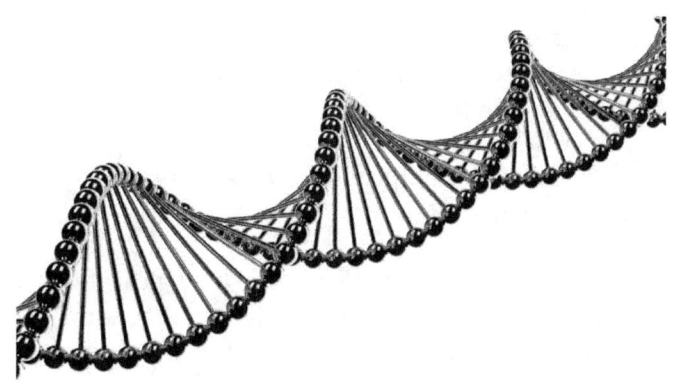

# Irreducible Complexity

We now turn from looking at evolution in relation to whole species of animals, to examining what would need to have happened at a cellular and molecular level for evolution to have taken place. For all the species alive today to have evolved from simple, single-celled life-forms, there would have needed to be unthinkably huge amounts of new complex genetic information created at every step. In fact, as you are about to see, even the simplest of living cells in the supposed very beginning would have required enormous amounts of complex micro-biological structures and genetic encoding to be created. To date, the theory of evolution has not been able to provide a viable means whereby this original complexity could have come about through natural means. Central to this dilemma, is the concept of irreducible complexity.

# Irreducible Complexity

Irreducible Complexity refers to the concept that even the smallest components of biological life are unimaginably complex, and cannot have arisen by chance. An irreducibly complex entity (such as a molecular machine) is one which is comprised of multiple interdependent parts, all of which are essential for the system to function effectively and, therefore, needed to come into existence simultaneously in order for the entity to be viable in the first place.

The concept of irreducible complexity, as an argument against evolution, has two fundamental premises.

**Firstly**, that some of the simplest elements of biological life are extremely complex, comprising multiple interdependent components, all of which are essential for the entity to function effectively. The entity cannot be simplified or reduced further without rendering it unviable.

**Secondly**, the multiple interdependent components of a biological entity could not have evolved by gradual, progressive construction and accumulation, because the entity would not function and would be unviable until all the necessary components were in place. All of the interdependent components had to come into existence simultaneously in order for the entity to exist, function effectively and be able to replicate and pass on its biological and genetic components to the next generation.

## IRREDUCIBLY COMPLEX MOLECULAR MACHINES

Life at the cellular and molecular level is unbelievably complex. Individual cells are bursting with tiny molecular robot-like "machines" which carry out very specific functions.

One example is the flagella of a single-celled bacterium, which is the bacteria's tiny tail that spins rapidly like a propeller, to give the bacterium its mobility. Electron microscopes reveal the remarkably detailed design of the flagellum, containing 40 independent, moving parts, precisely

## Irreducible Complexity

engineered to fit together to make the "propeller" work. The image below shows a diagram of the flagellum and its components, with an electron microscope image of a flagellum on the right:

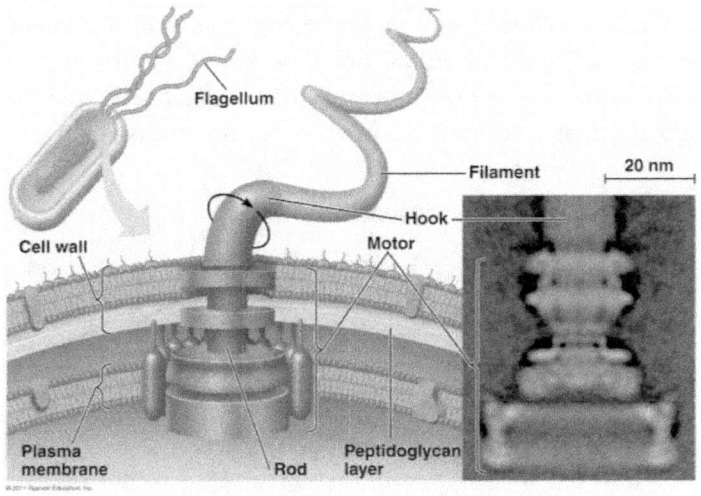

Bacterial Flagellum. (biologicalexceptions.blogspot.com)

A bacterial flagellum is an extremely efficient machine, rotating at 100 turns per second under normal conditions and increasing to 300 turns per second for short bursts.[1] Each of the 40 individual components of the flagellum is made of a specific protein, in a specific shape, precisely engineered to carry out a specific function that makes an essential contribution to the functioning of the flagellum. Electron microscopes reveal that the flagellum has a finely designed motor, with meshing cogs, a drive shaft, various connectors and the propeller itself. Without any one of its 40 individual components, the flagellum simply would not work. The flagellum is irreducibly complex; it cannot be reduced without losing its function.

Dr. Michael Behe, Professor of Biochemistry at Lehigh University, Pennsylvania, is a strong proponent of irreducible complexity as a refutation of evolution. He explains irreducible complexity by using a

# Irreducible Complexity

simple analogy. A spring and catch mouse trap contains five interdependent parts: a wooden platform, the spring, the hammer (which pins the mouse against the wooden base), the holding bar and a catch. Each of these components is absolutely essential for the mouse trap to function. Without any one of these, the mouse trap would not simply be less effective, it would not function at all! The mouse trap is irreducibly complex. Each of the components must exist and be in their correct positions simultaneously, in order for the mouse trap to function.

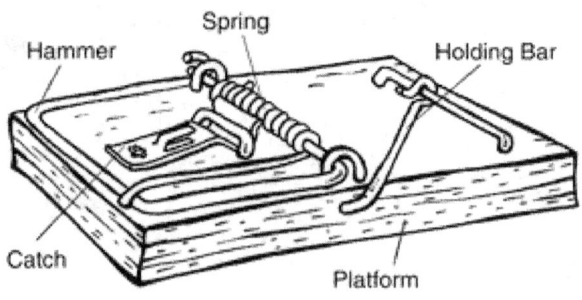

The Irreducibly Complex Mousetrap

In the same way, each of the 40 components of the bacterial flagellum need to be in place for the flagellum to work. Without all 40 components simultaneously existing, the flagellum would not work, the bacteria would not be able to move, and it would die without being able to replicate. This is strong evidence that the flagellum could not have formed gradually, through a succession of incremental modifications over a series of generations, because a fully functioning, complete flagellum is absolutely essential for a bacterium cell to survive and replicate. This is evident observationally, because individual bacterium that have dysfunctional or non-functioning flagellum do not survive and cannot replicate.

This contradicts evolution, which purports that all the biological components of life evolved gradually, in small incremental steps. It proposes that new biological structures such as the flagellum were built very gradually over many thousands of successive generations of the cell, with increasingly complex components being added until the structure was finally complete and functional. There are three problems with this, however.

***Firstly***, the bacterium would have no reason to retain non-functioning components in subsequent generations, because such components would be of no value to the bacterium until the whole structure is complete.

***Secondly***, as I have already explained, a bacterium without only a half-built (or one-fortieth built!) flagellum could not survive, and would therefore not be able to replicate in order to pass on new components to subsequent generations.

***Thirdly***, the creation of the new structural components necessary to build the flagellum would require large amounts of completely new DNA, and there is no reasonable explanation for how this new DNA could come into existence by random natural processes.

The bacterial flagellum is an irreducibly complex molecular machine that could not have been produced by gradual, incremental stages. All 40 precisely engineered components must have been present from the very beginning in order for the bacteria to be a viable organism, capable of reproduction. Evolution does not have an adequate explanation for this. The conundrum that this poses is occasionally admitted by evolutionists. For example, Drs. H. L. True and S. L. Lindquist, in a paper published in September 2000 Nature magazine, commented:

> '*A major enigma in evolutionary biology is that new forms or functions often require the concerted efforts of several independent genetic changes. It is unclear how such changes might accumulate when they are likely to be deleterious individually and be lost by selective pressure*"[2]

# Irreducible Complexity

However, the bacterial flagellum is not the only irreducible complex molecular machine. The micro-biological world contains many hundreds or possibly thousands of irreducibly complex machines, some of them containing many more that 40 interdependent moving parts. For example, the spliceosome - a molecular machine within cells that orchestrates the reading of sections of genes - contains over 300 separate components.[3] This brings us to a discussion of the cell itself.

## THE EXTRAORDINARY COMPLEXITY OF CELLS

A single living cell was once thought to be extremely simple; a mere blob of protoplasm surrounded by a cell wall. Scientists in the 19th century supposed that such a simple building block of life could easily have sprung into existence through the chance combinations of chemicals in the alleged distant primordial past. However, the development of the electron microscope, together with advances in the field of genetics, have completely overturned these naive assumptions. Even the simplest of cells is anything but simple!

As new technology has enabled scientists to explore the micro-biological dimension, a stunning new world of complex, inter-dependant molecular machines and biological processes has revealed itself. Simple cells are now understood to be extraordinarily sophisticated biological factories, teeming with hundreds of tiny molecular machines scurrying throughout the cell, carrying out hundreds of specialised tasks, all of which are essential for the cell to function effectively and remain viable. Some molecular machines transport nutrients from outside the cell, through the cell wall into the cell. Some carry nutrients throughout the cell. Some transport waste back to the cell wall. Some open up chemical holes in the cell wall to allow waste to be evacuated. Some repair other machines. Some build new machines. Some build the components of new cells. The ATP synthase is an example of one the hundreds of molecular machines within a living cell that is essential for the cell's viability.

# Irreducible Complexity

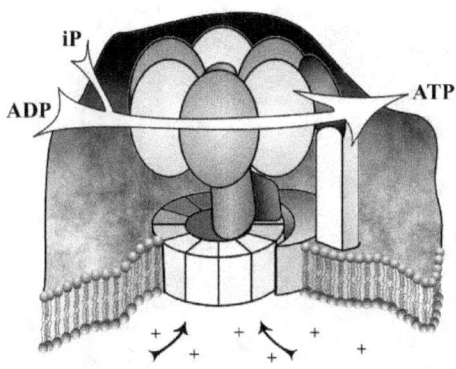

The ATP synthase (pictured above) is one of the more spectacular molecular machines that populate living cells. Respected biologist, Alex Williams, explains:

> *"It is a proton-powered motor that produces the universal energy molecule ATP (adenosine tri-phosphate). When the motor spins one way, it takes energy from digested food and converts it into the high-energy ATP, and when the motor spins the other way, it breaks down the ATP in such a way that its energy is available for use by other metabolic processes."*[4]

The ATP synthase machine is merely one of many hundreds of irreducibly complex molecular machines within a single living cell. A vast multitude of these irreducible complexity machines are required to sustain human life at the cellular level, making the possibility of evolution logically and mathematically unsustainable.

Harvard University Biology Department have produced a fascinating computer enhanced video of the inside of a single human cell, depicting the many hundreds of molecular machines moving and working together:[5]

# Irreducible Complexity

Screen shot from "The Inner Life of a Cell", Harvard University

The screen shot on the previous page is a computer-generated model of the inside of a cell, showing the spherical nucleus of the cell surrounded by the intra-cellular structures and molecular "highways". Travelling along these highways, hundreds of molecular machines can be seen moving throughout the cell, carrying out their vital functions. The "simple" cell is far from simple. It is an extraordinarily complex factory of hundreds of interdependent molecular machines.

**The Irreducible Complexity of a Living Cell**

Many scientists, particularly evolutionists, maintain that the complexity of the cell does not fall within the precise definition of *irreducible* complexity. However, as the extraordinary extent of the cell's complexity becomes increasingly apparent, some respected scientists are voicing an alternate viewpoint. For example, Dr. Tas Walker from Creation Ministries International asserts:

> *"In fact the whole structure of the living cell points to it being <u>irreducibly complex</u>. Furthermore, every biological pathway is <u>irreducibly complex</u>, requiring multiple steps to produce the required biochemical output. If one of the steps does not work, the product will not be produced."*[6]

# Irreducible Complexity

Similarly, Dr Jonathan Sarfati, in chapter 10 of his 2013 book, "*Refuting Evolution 2*", cites three examples of irreducible complexity; the eye, <u>the complex cell</u> and the bacterial flagellum.[7] Responding to evolutionist claims that a cell is not irreducibly complex, Dr. Sarfati states:

> "*Living things have fantastically intricate features - at the anatomical, cellular and molecular level - that <u>could not function if they were any less complex</u> or sophisticated.*"[8]

Irreducible complexity is established if a biological entity cannot be further reduced without destroying its viability. In other words, a biological entity is irreducibly complex if all of its components are essential to its survival and proper functioning, and must have been present within it from the very beginning. In the case of a cell, every one of its hundreds of components is absolutely essential for its survival.

**Firstly**, there is the vast army of molecular machines that carries out the many life-sustaining processes within the cell. There are molecular machines for converting energy stores to usable nutrients, for transporting nutrients around the cell, for removing waste, for repairing damaged molecular machines, for breaking down dysfunctional molecular machines and recycling their parts.[9] The cell is literally swarming with teams of specialised molecular machines! For example, there is the extraordinary assembly line of molecular machines within the nucleus, at the heart of the cell, responsible for producing new proteins:

- Molecular machines to uncoil a strand of DNA ready for transcription

- RNA polymerase machines to transcribe a section of DNA into messenger RNA (mRNA)

- Molecular machines to cut and splice sections of the mRNA

- Molecular machines to transport the mRNA outside the nucleus into the cytoplasm of the cell

- A molecular machine known as ribosome which binds to the mRNA and reads the code in the mRNA to produce a chain made up amino acids

- Transfer RNA molecular machines to link new amino acids to the mRNA chain in the correct order (thus forming a new protein),

- Molecular machines to then fold the new protein into its correct shape,

- Molecular machines to carry the new protein to the site in the cell where it is needed.

All of these machines are essential to the proper functioning and survival of the cell, and without all of them working together, the cell would die. In this sense, the humble cell can be said to be functionally irreducibly complex. Its viability (survival and ability to replicate) is absolutely dependent upon the simultaneous co-existence of this vast army of interdependent machines. Furthermore, the cell could not possibly have developed this army of molecular factory workers gradually and incrementally over time, because they are interdependent.

**Secondly**, there are the cell's impressively complex physical structures. Far from being a blob of shapeless protoplasm, a cell is an intricately designed entity comprised of many complex structural elements. Consider the following diagram of the structural components of a cell:

## Irreducible Complexity

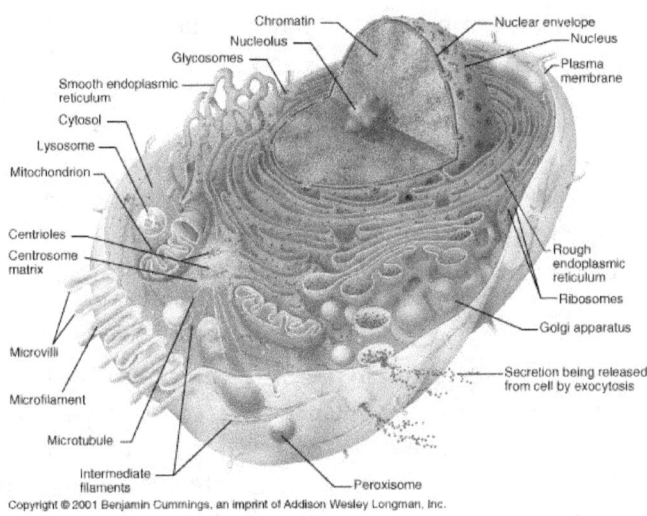

A cell has approximately 30 different structural elements, all playing a vital role in its viability. If the cell had no mitochondria for converting nutrients into energy molecules, it would die. If it had no cell wall, it would die. If it had no nucleus with its repository of DNA to provide the cell with its operating instructions, it would die. In fact, remove any of the structural elements from a functioning cell, and it will cease to be viable and, in almost all most cases, will die without replicating. All of these structural components had to be in place from the very beginning; they all had to come into existence simultaneously in order for the first cell exist as a living, reproducing biological entity. In this sense, the cell can be said to be structurally irreducibly complex.

As Dr. Tas Walker states:

> "The whole structure of the living cell points to it being irreducibly complex."[10]

I am aware that there are staunch evolutionists who would argue against the cell's irreducible complexity. They theorise that the cell's structures and processes could have evolved gradually and incrementally over time, and that the original, primordial cells may have been much simpler. There are two objections to this line of reasoning:

**Firstly**, this argument is purely speculative. It arises from the imaginations of evolutionists who are envisioning the existence of simpler primordial cells. No such cells can be observed today. The argument for the cell's irreducible complexity is based upon current observable evidence, whereas the evolutionist's contrary argument is based upon pure speculation.

**Secondly**, evolutionists have been unable to propose a verifiable means by which even one of these irreducibly complex molecular machines or complex physical structures within a cell could have come into existence via natural means. Ongoing attempts to create even a single protein in laboratories with the latest technologies, have completely failed. Yet, as Dr. John Ashton points out, a single cell has 2.4 million protein molecules made up of approximately 4,000 different types of proteins.[11] And all of these proteins needed to come into existence simultaneously in order for the single cell to be born.

Dr. Thomas Heinze, professor of organic chemistry at Friedrich Schiller University, Jena, in his article, "*Did God Create Life? Ask a Protein*"[12], states:

> "*Proteins are so complex they will not form anywhere in nature except in living cells. Inside cells, the directions for protein construction are already contained in the DNA. Then, if a protein is to perform its task, its production must be carefully regulated, but even then, it will not function unless it also has the correct address tag and is properly folded. All these systems would have to have been in place or the 'first cell' could not function. These systems, however, are just the tip of the iceberg. I chose them to illustrate <u>the many coordinated systems that would have to have been present before the first cell would work.</u>*

# Irreducible Complexity

> *The teaching that the first cell spontaneously popped into being without the involvement of the Creator has its basis in the pre-scientific myth that single-celled creatures were simple. It obviously does not stand up under today's knowledge that a cell's DNA, RNA, membranes and proteins are extremely hard to make, and when proteins are made, they must be properly folded, addressed, and turned on and off at just the right times. None of these brilliant solutions could invent itself, yet <u>no 'first cell' could exist without all of them</u>. They could not have happened without a very intelligent Creator."*

Dr. Heinze's testimony concerning the many complex biochemical intra-cellular systems that had to exist simultaneously in order for the first cell to come into existence is strong evidence for the irreducible complexity of the cell - complexity that points strongly to the intervention of a Grand Designer.

**Autopoiesis**

In response to the reluctance of some within the scientific community to classify the cell as irreducibly complex, as well as a variety of refutations from evolutionists disputing the concept of irreducible complexity itself, some creationists have pointed to the concept of autopoiesis. Autopoiesis refers to the self-perpetuating nature of the basic components of all living organisms. It describes the fact that cells are self-maintaining, self-organising, self-repairing, self-replicating and self-perpetuating, and that the origin of such inherent "self-life" lies beyond the reach of naturalistic explanations. Biologist, Alexander Williams, believes autopoiesis makes a stronger case for intelligent design than irreducible complexity because it is inherent in *all* biological life. Furthermore, he explains that:

> *"Autopoiesis provides a compelling case for intelligent design in three stages: (i) autopoiesis is universal in all living things, which makes it a pre-requisite for life, not an end product of natural*

> *selection; (ii) ... <u>autopoiesis is not reducible</u> to the laws of physics and chemistry [cannot be explained by mere naturalistic processes]; and (iii) there is an unbridgeable abyss between the dirty [inefficient], mass-action chemistry of the natural environmental and the perfectly-pure, single-molecule precision of [intra-cellular] biochemistry."[13]*

Williams refers to the "*machine-like structures that exist in living organisms*" as "<u>*irreducible structures*</u>" which "<u>*cannot be reduced*</u> *to the properties of the carbon, hydrogen, oxygen, nitrogen, phosphorus, sulphur and trace elements that they are made of.*"[14] In this sense, he uses the concept of autopoiesis to define irreducible complexity more broadly, as the irreducible self-sustaining processes of the living cell that are without naturalistic explanation, because they are greater than the sum of their chemical components. In other words, there is a life-perpetuating essence of the cell that defies chemical explanation. Williams states:

> "*Autopoiesis is a unique and amazing property of life - there is nothing else like it in the known universe. It is made up of a hierarchy of irreducibly structured levels.*"[15]

By "irreducibly structured levels", Williams is referring, in part, to the many complex, interdependent biological systems within the cell that contribute to its life-perpetuating essence. The lack of an adequate naturalistic explanation for how all these interdependent systems could have developed *simultaneously* is cautiously acknowledged evolutionists. For example, Drs. M. W. Kirschner and J. C. Gerhart, in their book, "*The Plausibility of Life: Resolving Darwin's Dilemma*", concede:

> "*Core processes [within the cell] may have <u>emerged together as a suite</u>, for we no of no organism today that lacks any part of that suite ... The most obscure origination of a core process is the creation of the first prokaryotic [simple] cell. The novelty and*

*complexity of the cell is so far beyond anything inanimate in the world of today that we are left baffled."*[16]

**Intelligent Design Advocates**

The growing concept of the irreducibly complex, factory-like nature of the single cell led to one of the most famous scientific conversions of the last century. Dr. Dean Kenyon, Professor Emeritus of Biology at San Francisco State University, was once a convinced evolutionist who co-authored the evolutionary biology textbook, *"Biochemical Predestination"*.[17] In the late 1970s, however, he became increasingly aware of a variety of evidence contradicting evolution, including the inexplicable irreducible complexity of the single cell. Dr. Kenyon became convinced that the theory of evolution could not explain how such an irreducibly complex system could have evolved, and he became an outspoken advocate of intelligent design and creationism. He began teaching creationism in his biology courses at university, and he became a strong proponent of creationism in various public debates and court cases.[18]

Dr. Kenyon published many subsequent academic papers and books, repudiating his previously held position and advocating strong evidence for intelligent design.[19] This intelligence transcends the ability of the most learned and competent scientists to design or manufacture a comparable machine, even if they were able to manufacture each component independently and then assemble later. Science, with its ever-increasing knowledge and technology, has not even come close to emulating what nature is supposed to have done by chance, random processes!

In an interview in the 2010 Illustra Media video, *"Unlocking the Mystery of Life"*, he comments on the profound impact that the irreducible complexity of the single cell had upon him,stating:

> *"This is absolutely mind boggling to perceive at this scale of size, with such finely tuned apparatus. ... We see the details of an*

> *immensely complex molecular realm of genetic information processing and it is exactly this new realm of molecular genetics where we see the most compelling evidence of design on the earth. Nothing short of an intelligence could have created this intricate cellular apparatus."[20]*

Dr. Michael Behe's book, *"Darwin's Black Box: The Biochemical Challenge to Evolution"*[21] was a ground-breaking academic work arguing for the impossibility of evolution to create irreducibly complex systems. In it he explains the main argument against evolution:

> *"By irreducibly complex I mean a single system composed of several well-matched, interacting parts that contribute to the basic function, wherein the removal of any one of the parts causes the system to effectively cease functioning. An irreducibly complex system cannot be produced directly (that is, by continuously improving the initial function, which continues to work by the same mechanism) by slight, successive modifications of a precursor system, because any precursor to an irreducibly complex system that is missing a part is by definition nonfunctional. An irreducibly complex biological system, if there is such a thing, would be a powerful challenge to Darwinian evolution."[22]*

Since it's publication, *"Darwin's Black Box"* has elicited furious debate from evolutionists who claim to be able to refute his arguments, but their arguments are theoretical and semantical. To date, no evolutionist has been able to offer a practical, observable, verifiable process whereby a prolonged series of random, successive, *non-functional* modifications can lead to a fully-functioning irreducibly complex system. In response to his critics, Dr. Behe and other intelligent design advocates repeatedly point out that the gradual, successive, development of non-functioning components would be of no advantage to a cell until all the components are present; and without all the components present, the cell simply could not survive and replicate in order to pass on those partial modifications![23] Furthermore, each successive component of a molecular machine would require huge amounts of new DNA information, and, as we will investigate in the next chapter, evolution has absolutely no explanation as to how such new information can appear out of thin air!

Dr. Behe is one of many scientists today who are proclaiming the impossibility of evolution in the light of irreducible complexity. An article published on the Intelligent Design and Evolution Awareness Centre (IDEA) website, examines, among other examples of irreducible complexity, the cilia on the outside of single-celled organisms; hair-like filaments used to propel a cell through fluid. The article states:

> "Evolution simply cannot produce complex structures in a single generation as would be required for the formation of irreducibly complex systems. To imagine that a chance set of mutations would produce all 200 proteins required for cilia function in a single generation stretches the imagination beyond the breaking point. And yet, producing one or a few of these proteins at a time, in standard Darwinian fashion, would convey no survival advantage because those few proteins would have no function- indeed, they would constitute a waste of energy for the cell to even produce. Darwin recognized this as a potent threat to his theory of evolution - the issue that could completely disprove his idea. So the question must be raised: Has Darwin's theory of evolution "absolutely broken down?" According to Michael Behe, the answer is a resounding 'yes'."[24]

## DNA

Perhaps the most stunning example of the extraordinary complexity of the microbiological world is the existence of DNA. It may be helpful to have a basic understanding of the nature of DNA (deoxyribonucleic acid). Below is a basic introduction to DNA, sourced from the U.S. National Library of Medicine's Genetics Home Reference website. If you find it difficult to understand, don't be too perturbed, as a very simple explanation is provided immediately following this. But for those readers who are interested in the more technical aspects of DNA, the following explanation will be helpful:

> "DNA, or deoxyribonucleic acid, is the hereditary material in humans and almost all other organisms. Nearly every cell in a person's body has the same DNA. Most DNA is located in the cell nucleus (where it is called nuclear DNA), but a small amount of DNA can also be found in the mitochondria (where it is called mitochondrial DNA or mtDNA). Mitochondria are structures within cells that convert the energy from food into a form that cells can use.
>
> The information in DNA is stored as a code made up of four chemical bases: adenine (A), guanine (G), cytosine (C), and thymine (T). Human DNA consists of about 3 billion bases, and more than 99 percent of those bases are the same in all people. The order, or sequence, of these bases determines the information available for building and maintaining an organism, similar to the way in which letters of the alphabet appear in a certain order to form words and sentences.
>
> DNA bases pair up with each other, A with T and C with G, to form units called base pairs. Each base is also attached to a sugar molecule and a phosphate molecule. Together, a base, sugar, and phosphate are called a nucleotide. Nucleotides are arranged in two long strands that form a spiral called a double helix. The structure of the double helix is somewhat like a ladder, with the base pairs forming the ladder's rungs and the sugar and phosphate molecules forming the vertical sidepieces of the ladder.
>
> An important property of DNA is that it can replicate, or make copies of itself. Each strand of DNA in the double helix can serve as a pattern for duplicating the sequence of bases. This is critical

*when cells divide because each new cell needs to have an exact copy of the DNA present in the old cell."*[25]

There! That wasn't so bad, was it!

So, what does all this mean? Let me break it down into a much simpler explanation.

DNA is the "computer coding" that resides within the nucleus of almost every cell in your body. It is the complete instruction manual of how to build every component of your body, how to repair it when bits break and how to ensure its ongoing proper functioning. It is a vastly complex piece of code. It contains 3.2 billion "letters" which are combined to spell out very specific genetic instructions, built into a biological construct in the shape of a spiralling double helix. To give you an idea of how much information is in human DNA, the Encyclopedia Britannica has 200 million letters across 32 volumes. Human DNA, by comparison, has 3.2 **billion** letters - that is approximately 16 sets of the 32 volume Encyclopedia Britannica! To put it another way, if DNA could be printed as words in a book, it would fill up 12,000 copies of the book you are currently holding in your hand (or looking at on your screen)!

A further startling fact (which is completely irrelevant but extremely fascinating) concerns the length of your DNA chain. The DNA in the nucleus of each cell is an extremely fine molecular structure, folded in upon itself, over and over again. If you could unfold your DNA into a straight line, the DNA in a **single cell** would be about 2 metres long (6 feet)[26] - although you would need an electron microscope to see it, because it would only be a single molecule in width. That is the length of the DNA in just **one** cell. Your body has about 10 **trillion** cells. If you could unfold all the DNA from every cell and place it end to end, your DNA would stretch out to 1.2 million kilometres (744 million miles)! That would stretch from the earth to the sun and back again, **4 times**! I told you it was interesting!

But it is also completely irrelevant, so forget I mentioned it.

What *is* relevant to a discussion of evolution, is the incredibly complex nature of DNA and the vast amount of information it contains. The theory of evolution has no satisfactory explanation for how that huge amount of complex coding got into our cells in the first place!

The Human Genome Project was an international research project aimed at mapping the exact sequence of base pairs (or "letters") within the entire human DNA. It was an enormous task and remains, to this day, the largest collaborative biological project ever undertaken. Starting in 1990, using the most sophisticated computers available and involving hundreds of geneticists from around the world, it took until 2003 until the entire DNA sequence was mapped.[27] It took **13 years**, with the best available computers running 24/7, to determine the exact order of all 3.2 billion base pairs within our DNA! That is how vast and complex the coding is within our DNA!

Furthermore, although we now know the exact order of all the "letters" (base pairs) within our DNA, scientists only understand the exact nature and function of **less than 1%** of those biological instructions.[28] We have mapped our DNA, but we have very little understanding of what most of it "says". It is a highly complex code. Even today, with our most sophisticated computers and technology, we have barely begun to understand the complex code of our DNA.

Where did this vast amount of complex information come from? Who wrote the code and placed it inside every cell of our body? The Bible points us towards an all-powerful God who designed and created us, and who inscribed this incredibly complex code into every cell of every living organism, but the evolutionary story has no explanation for this. No amount of random mixing of chemicals or stirring of the primordial swamp can conjure up a process whereby this vast amount of complex coding simply pops into existence.

## Irreducible Complexity

And in case you are thinking that the earliest, simple lifeforms would have needed a much smaller amount of DNA, consider the following. According to the U.S. National Science Foundation,[29] our human DNA contains about 23,000 genes (a gene is a section of DNA, containing a whole "page" or "chapter" of instructions regarding a particular function or organ). By comparison, a water flea (*Daphnia pulex*) has 31,000 genes![30] The humble earth worm has 20,000 genes![31] Even more astonishing is the complex DNA chain found within single-celled organisms. Compared to human DNA, which has 3.2 billion base pairs, the single-celled amoeba proteus has 290 billion base pairs and the single-celled amoeba dubia has 670 billion base pairs![32] In other words, amoeba DNA is **90 times** longer than human DNA and protozoa DNA is **209 times** longer! Scientists have very little understanding of why this is so,[33] but one thing it demonstrates very clearly is that there is no such thing as a "simple" organism.

Dr. Macki Giertyche, Professor Emeritus of Genetics, at Torun University, Poland, states:

> *"The science of genetics makes it clear that at no time in the past can there have been such a thing as a simple organism. All organisms, however primitive they may appear, are complex and bursting with information. And we know that this information must have been there from the very beginning. For example, the very complex DNA & RNA protein replicating system in the cell must have been perfect from the very beginning - if not, life could not exist. The only logical explanation is that this vast quantity of information came from intelligence. Every bacterium, every microscopic cell, is so precisely programmed that we have to assume that the information contained in them comes from an intelligence far beyond our own.... Evolutionists have no idea how this information system is produced."*[34]

Evolution teaches that life began with very simple, single-celled organisms and gradually evolved into more complex life-forms. But our recent knowledge of genetics reveals that even the smallest of organisms is incredibly complex; filled with a vast amount of intelligent, highly detailed coding that could not possibly have arisen by chance. No amount of lightning strikes into a primordial swamp could instantly (or ever!) create a series of biological instructions equating to, in the case of amoeba dubia, **15,000 sets** of the 32 volume Encyclopedia Britannica!

As Dr. John Ashton states, in his book, *"Evolution Impossible"*:

> *"To date, I have found no reputable published scientific paper that explains a proven mechanism for how this huge amount of highly complex genetic information could arise by chance. Nor can I find any scientific papers reporting the observation of new meaningful genetic information arising by chance."*[35]

What is also remarkable about DNA is the ability of individual cells to decode and read the portion of the DNA that is applicable to their own function, and ignore the rest. In the case of humans, almost every cell in our body has a complete strand of DNA stored in its nucleus- a complete instruction manual for the building, repairing and proper functioning of the whole human body. Yet, remarkably, each cell only accesses that part of the DNA that is applicable to itself, and disregards the rest. Although the world's most gifted scientists have only been able to decode less than 1% of our DNA (they have almost no idea what the rest of it means - what each segment is for), our cells can read it easily, and can extract the necessary instructions and follow them. Who programmed our cells to be able to do this? How does an eye cell know which part of the DNA to access and read? How does a liver cell know where to look in the instruction manual to repair itself? This extraordinary ability of our cells is another aspect of our biological sophistication and complexity that cannot be explained by random evolution.

All of the above demonstrates that DNA is both irreducibly complex and has a degree of specified complexity that cannot be explained by evolutionary processes. Let us now briefly examine these two concepts.

## DNA AND IRREDUCIBLE COMPLEXITY

In order for a cell to be alive and able to reproduce, it requires its complete set of DNA. This genetic coding cannot have accumulated over vast amounts of time, because without a full set of DNA, with a complete set of genes to regulate all of the cell's functions, the cell would not be alive and would not be able to replicate and pass on its genetic information to subsequent generations.

The minimum requirement for a single cell to be a living, functioning, reproducing entity, is a *full set* of genetic coding, with all the relevant genes, to be present *from the very beginning*. In this sense, the DNA within a cell is irreducibly complex. For example, the previously mentioned single-celled amoeba dubia needs the complete DNA of 670 billion base pairs to present in order for it to exist. An amoeba dubium with only 300 billion base pairs would be incomplete and unviable. It would not be a living amoeba at all, but merely a random collection of chemicals and molecules. The full strand of DNA had to come into existence *in its completed form* in order for any cell to be alive and able to reproduce. Gradual, incremental accumulation of DNA information is impossible without a living organism to reproduce and pass on that incremental information. And a living organism is not possible without a full set of DNA. In the words of Joseph Heller's famous book, it's a "Catch 22" situation![36]

A recent article on the website, Creation Science Hall of Fame, states that amoeba dubia contains enough information in its DNA chain, to fill the U.S. National Library of Congress 10 times over![37] How could that vast amount of genetic information have come into existence simultaneously and by chance?

Charles Darwin, of course, knew nothing of DNA when he published his evolutionary theory in 1859. DNA was not properly understood until 1953.[38] Had Darwin and his contemporaries understood the complex nature of DNA, they would most likely never have entertained the concept of life popping into existence by chance natural causes.

## DNA AND SPECIFIED COMPLEXITY

As well as irreducible complexity, DNA also demonstrates specified complexity which cannot be adequately explained by evolution. Specified complexity refers to the existence of a highly complex structure or set of information that is required to exist in a very specifically defined sequence in order to be viable. The longer and more complex the sequence of information, the less likely it is that it came about by chance processes.

### Statistical Impossibility

Some atheist scientists try to argue that the vast amount of information in the DNA strand could have developed by chance. They argue that a monkey tapping away at random on a computer keyboard would eventually correctly type not only words but also phrases, sentences, and even whole books.[39] So, given enough time, they argue that chance alone could produce the necessary genetic information for anything - whales, monkeys, humans.

Let us examine this hypothesis logically. As there are 101 keys on a typical computer keyboard, the chance of a monkey randomly typing the six-letter word *"peanut"* is 1/101 x 1/101 x 1/101 x 1/101 x 1/101 x 1/101, which equals 1 chance in 1.06 trillion. Typing at 3 letters per second, the monkey would have to type for 10,379 years to produce the word "peanut"! What about the phrase *"peanuts and computers"*? To correctly type these 21 letters and spaces by chance alone, it would take 1 billion monkeys typing randomly for one thousand million million million million million years! That is longer than even the most ardent evolutionists believe the universe has existed! But there aren't just 21 letters in our DNA; there are 3.2 billion base pairs! This shows how utterly ridiculous it is to believe that the vast amounts of information in DNA could have come into existence by random natural processes. And, as previously stated, for a cell to be a viable, living, functioning, reproducing entity, the **entire** DNA chain had to come into existence **simultaneously** and in **precisely** the correct sequence.

## CHARLES DARWIN'S FRANK ADMISSION

Even as Charles Darwin published his theory in 1859, he was acutely aware of its tenuous nature. In a statement on page 189 of his book, "*On the Origin of Species*", he made this frank admission:

> "If it could be demonstrated that any complex organ existed which could not possibly have been formed by numerous, successive, slight modifications, my theory would absolutely break down."[40]

Increasingly today, many scientists are claiming that we have now discovered numerous examples of the kind of irreducibly complex biological systems that Darwin stated could disprove his theory. Biochemist, Dr. Michael Denton, in his book, "*Evolution: A Theory in Crisis*", states:

> "We now know that there are in fact tens of thousands of irreducibly complex systems on the cellular level. Specified complexity pervades the microscopic biological world. Although the tiniest bacterial cells are incredibly small, weighing less than $10^{-12}$ grams, each is in effect a veritable micro-miniaturized factory containing thousands of exquisitely designed pieces of intricate molecular machinery, made up altogether of one hundred thousand million atoms, far more complicated than any machinery built by man and absolutely without parallel in the non-living world."[41]

Commenting on Darwin's frank admission regarding the possibility of an irreducibly complex system disproving his theory, Dr. Michael Behe states:

> *"As the number of unexplained, irreducibly complex biological systems increases, our confidence that Darwin's criterion of failure has been met skyrockets toward the maximum that science allows."*[42]

One suspects that if Charles Darwin was alive today, he would concede defeat, based upon his own criteria of failure. Many of his present-day supporters, however, appear much less open to evaluating the theory in the light of contrary evidence than Darwin was, himself.

## CONCLUSION

Evolution teaches that biological life gradually developed from the simple to the complex. But we now know that there is no such thing as a simple organism. Even the smallest, single-celled lifeform is irreducibly complex. The DNA in the nucleus of almost every living cell contains inconceivably vast amounts of detailed, coded information, sufficient to fill the largest of libraries multiple times, and cannot be explained by gradual, random processes. Furthermore, we now understand that each cell is an inconceivably complex biological factory, consisting of hundreds of highly specialised, inter-dependent molecular machines. Each of these molecular machines is, itself, comprised of multiple inter-dependent parts, made of precisely engineered proteins manufactured by other molecular machines within the cell. Every one of these components is absolutely essential to the viability of the cell, and all elements of this incredible cellular factory needed to come into existence simultaneously in order for the cell to survive and replicate. This kind of irreducible complexity, at the smallest, most fundamental biological level, must cause us to seriously consider the possibility that Darwin's candidly admitted criterion of failure has been met, and reveals the complete inability of the theory of evolution to provide an explanation for the origin of life.

---

## ENDNOTES - Chapter 7

---

[1] http://book.bionumbers.org/what-is-the-frequency-of-rotary-molecular-motors/

[2] True, H.L. and Lindquist, S.L., A yeast prion provides a mechanism for genetic variation and phenotypic diversity, *Nature* **407**(6803):477–483, 28 September 2000, p. 477.

[3] Alex Williams, "Life's Irreducible Structure - Part 1: Autopoiesis", 2007, in Journal of Creation 21(2):109-115

[4] Alex Williams, "Life's Irreducible Structure - Part 1: Autopoiesis", 2007, in Journal of Creation 21(2):109-115

[5] https://www.youtube.com/watch?v=wJyUtbn0O5Y

[6] Tas Walker, article entitled, "Irreducible Complexity and Culdesacs", 17 December 2016, Creation Ministries International.

[7] Jonathan Sarfati, "Refuting Evolution 2", 2013, Creation Book Publishers, Powder Sprongs, GA, Ch. 10, point 14.

[8] Jonathan Sarfati, "Refuting Evolution 2", 2013, Creation Book Publishers, Powder Sprongs, GA, Ch. 10, point 14.

[9] Alex Williams, "Life's Irreducible Structure - Part 1: Autopoiesis", 2007, in Journal of Creation 21(2):109-115

[10] Tas Walker, "Irreducible Complexity and Cul-De Sacs", https://creation.com/irreducible-complexity-and-cul-de-sacs, 17 Dec 2016

[11] John F. Ashton. "Evolution Impossible", Green Forest, AR, Masterbooks, 2013, p.40

[12] Thomas Heinze, "Did God Create Life? Ask a Protein.", https://creation.com/did-god-create-life-ask-a-protein

[13] Alex Williams, "Life's Irreducible Structure - Part 1: Autopoiesis", 2007, in Journal of Creation 21(2):109-115

[14] Alex Williams, "Life's Irreducible Structure - Part 1: Autopoiesis", 2007, in Journal of Creation 21(2):109-115

[15] Alex Williams, "Life's Irreducible Structure - Part 1: Autopoiesis", 2007, in Journal of Creation 21(2):109-115

[16] M.W. Kirschner and J. C. Gerhart, "The Plausibility of Life: Resolving Darwin's Dilemms", Yale University Press, New Haven, CT, 2005, pp. 253-256

[17] Kenyon DH, Steinman G. Biochemical Predestination. McGraw Hill Text (January, 1969) ISBN 0-07-034126-5.

[18] https://en.wikipedia.org/wiki/Dean_H._Kenyon

[19] https://en.wikipedia.org/wiki/Dean_H._Kenyon

[20] "Unlocking the Mystery of Life", Illustra Media, 2010

[21] Michael Behe, "Darwin's Black Box: The Biochemical Challenge to Evolution", Free press, New York, 2006

[22] Michael Behe, "Darwin's Black Box: The Biochemical Challenge to Evolution", Free press, New York, 2006, p.39

[23] https://evolutionnews.org/2011/03/michael_behes_critics_make_dar/ See also: http://www.talkorigins.org/faqs/behe.html

[24] http://www.ideacenter.org/contentmgr/showdetails.php/id/840

[25] https://ghr.nlm.nih.gov/primer/basics/dna

[26] McGraw Hill Encyclopedia of Science and Technology. New York: McGraw Hill, 1997.

[27] https://www.genome.gov/12011238/an-overview-of-the-human-genome-project/ See also, https://en.wikipedia.org/wiki/Human_Genome_Project.

[28] See the Harvard University article, "The 99% of the Human Genome", http://sitn.hms.harvard.edu/flash/2012/issue127a/. See

[29] https://www.nsf.gov/news/news_summ.jsp?cntn_id=118530

[30] https://www.nsf.gov/news/news_summ.jsp?cntn_id=118530

[31] https://www.ncbi.nlm.nih.gov/pmc/articles/PMC138976/

[32] http://www.genomenewsnetwork.org/articles/02_01/Sizing_genomes.shtml

[33] http://www.genomenewsnetwork.org/articles/02_01/Sizing_genomes.shtml

[34] "Unlocking the Mystery of Life" DVD, Illustra Media, 2010

[35] John F. Ashton. "Evolution Impossible", Green Forest, AR, Masterbooks, 2013, p.22.

[36] Joseph Heller, "Catch 22", 1962, Random House, London

[37] http://creationsciencehalloffame.org/defenses/design/irreducible-complexity/

[38] https://profiles.nlm.nih.gov/SC/Views/Exhibit/narrative/doublehelix.html

[39] https://www.telegraph.co.uk/technology/news/8789894/Monkeys-at-typewriters-close-to-reproducing-Shakespeare.html

[40] Charles Darwin, "The Origin of Species", London, John Murray, 1859, p.189

[41] Michael Denton, "Evolution: A Theory in Crisis," 1986, p. 250.

[42] Michael Behe, "Darwin's Black Box: The Biochemical Challenge to Evolution", Free press, New York, 2006

# Impossible Genetics

Chapter 8

# Impossible Genetics

Nothing demonstrates the fundamental impossibility of evolution more clearly than the science of genetics. The latter half of the 20th century saw an explosion of knowledge in this field which has continued into the 21st century. As scientific understanding has increased regarding the complex structure of DNA, the function of genes and the processes of genetic replication and recombination, there has been a corresponding rapid deterioration of confidence in the plausibility of evolution. To put it simply, as we learn more about genetics, the impossibility of evolution becomes increasingly apparent. As Dr. Ken Ham states:

*"If they had known about genetics in Darwin's day, the theory of evolution would never have gotten off the ground."*[1]

So, let us briefly examine the nature of these problems, and investigate why an increasing number of respected geneticists are convinced of evolution's impossibility.

## CHARLES DARWIN and NATURAL SELECTION

Darwin's theory proposed that all biological life on earth evolved gradually, over time, from a single, simple biological ancestor. Darwin asserted that all biological species, from worms to humans, are related to each other, and developed along divergent lines through a process he termed *"natural selection"*. He proposed that environmental factors produced slight modifications within successive generations of a species, which accumulated over time, until eventually the changes were so significant that an entirely new species emerged. Bacteria grew eyes, gills, fins and internal organs and turned into fish. Fish grew legs and air-breathing lungs and emerged from the sea. The newly emerged aquatic life gradually morphed into ants, beetles, mice, dogs and elephants. Reptiles grew wings and became birds. Primates climbed down from the trees, grew more erect, developed vastly more complex brains, and became humans.

Darwin's concept of natural selection rests upon an assortment of simple observations that he made in the course of his travels. These were previously mentioned in Chapter 2, *"Darwin's Imaginative Theory"*, but it is relevant to list his observations again here:

- Variations of features within the same species of grass and plants.

- The shared basic structure of all flowers (all flowers have petals, sepals, stamens and pistils).

- Variations of beak length in finches on the Galapagos Archipelago.

- The discovery, on the islands of Madeira, of wingless beetles living alongside winged beetles.

- Similarities in physical structures among widely different species. (For example the similarity in limbs between many different species).

Based upon these simple observations, Charles Darwin proposed that, given sufficient time, natural selection could, gradually and incrementally, bring about the huge biological changes necessary for all the various species to develop.

But does the science of genetics substantiate this belief? Can the processes which create variations of beak length in finches explain how a single-celled bacterium can eventually turn into a professor of astrophysics? Can a frog turn into a prince? Can adaptation and variation within species account for the major changes required for evolution to take place (one species evolving into a completely different species)?

The answer that the science of genetics now provides, is a resounding *"No!"*.

## ADAPTATION & VARIATION

The fact that variations occur within a species is not in question. The kind of variations that Darwin observed within finches are evident within all species of biological life. Differences in size, shape and functional abilities are observably abundant within every species on our planet. Furthermore, these variations are not static, but can be observed to develop over a relatively short time-period in response to a variety of factors, including changes to the immediate environment.

In order to have a basic understanding of how adaptation and variation works, there is some technical information that is helpful. I will try to keep it as simple as possible! Adaptation is basically a change in gene frequency or gene dominance within a population. A gene is a section of DNA which has specific coding for a particular biological function or appearance. Humans have about 23,000 genes in their DNA. If the 3.2 billion base pairs that comprise our DNA are like letters spelling out words, then genes can be compared to self-contained chapters or clusters of words within the DNA. Each of these chapters (genes) contains very specific instructions for the formation and functioning of a particular component or process within the body. There are genes that determine every aspect of who you are; your eye colour, your height, your skin colour, your propensity to store fat, the shape of your nose, and so on. You actually have two sets of every gene, one set inherited from your father and one set from your mother, and these are tucked away in twin sets of 23 chromosomes, inside every one of your cells. In the midst of all these genes, some are dominant, some are recessive, and some are dormant (not yet switched on or active).

Now let me make this practical. In his book, "Evolution Impossible", Dr. John Ashton uses the example of adaptation and variation among mice.[2]

> *"If a mouse population that is carrying genes for both light and dark fur moves to a light-coloured sandy area where owls can see and catch the dark mice more easily, after a while there will be fewer dark mice to breed. As the light-coloured mice continue to breed, fewer and fewer of them will carry the genes for dark fur, so natural selection for light-coloured mice will have occurred. However, some mice may still be carrying the genes for dark fur, and if some of the light-coloured mice migrate to a dark soil area after breeding for a while, some dark offspring may now be produced. These now have a better chance of surviving the predator owls, creating a situation where light-coloured mice evolve into dark coloured mice."*[3]

The above example demonstrates how environmental factors can lead to adaptation and variation within a species. The original mice have within their DNA the genetic information for a variety of fur colours. Sexual reproduction will tend to randomly produce a variety of different colour-toned offspring; some will be lighter in colour and some will be darker. This is because the inheritance of genes from both parents and the random blending of those genes in the various offspring will make either the lighter or darker gene more dominant. If no environmental factors impinge upon the mice population, they will continue to produce offspring of varying colour tones, as the genes are randomly mixed and re-mixed through sexual reproduction.

But when environmental conditions favour a particular colour, not many of the alternate-coloured mice will live long enough to breed. In fact, if these environmental condition persist for sufficient time, the point can be reached where the genetic information for the alternate colouring is completely lost from the gene pool. In such a case, this particular mice population has permanently become a dark-furred variety. Hence, a new sub-species or variety of mice has been created.

However, it is vital to understand that in this example, the creation of a new sub-species is the result of the *loss* of genetic information. The dark mice have *lost* the genetic information for light colouring. **No new genetic information has been added to their DNA.** Almost all adaptation and variation are of this kind. The appearance and characteristics of a population become permanently altered because the genetic information for alternate characteristics have been lost or become dormant (present within the DNA, but switched off).

The recent and rapidly developing field of epigenetics reveals how environmental factors can produce rapid changes within a population by switching genes on or off. For example, in recent experiments on agouti mice, researchers found that simply by changing the diet of the mice they could switch off a certain gene that regulated size and colour.[4] When the gene is active, the mice are normally obese and yellow, but by feeding the mice a diet high in vitamin B12, the relevant gene was switched off and the resulting offspring were small and brown.[5]

Consider the various breeds of dog, as another example. Geneticists say that it is probable that, originally, there was just one, generic breed of ancestral wolf. That breed would have had a complete set of "doggy DNA" with information within its DNA that could have produced a wide variety of characteristics: long hair / short hair; curly hair /straight hair; big ears / small ears; long tail / short tail; long legs /short legs, etc. Random breeding through sexual reproduction would have produced very minor variations in these and many other characteristics within a single litter of puppies. What is remarkable about adaptation and variation, is that if dogs with certain characteristics breed with other dogs with the same characteristics, the genes for those characteristics are strengthened in their offspring, while the genes for the alternate version of that characteristic are eventually lost. Eventually, if dogs continue to breed with other dogs with similar characteristics, fairly rapid transformation can occur. This process can be accelerated even more by "artificial selection"; a process whereby animals with traits desirable to human breeders are systematically selected and bred together. This is the process that has been used to create many new breeds of dog over the last few centuries.

Once again, it must be stressed that the creation of a new breed, in this way, results from the **loss** of certain genetic information and the **reinforcing** of other genetic information that was already present. No **new** genetic information has been added. A dog breed with short, curly hair has simply lost the genes for long straight hair. A dog breed with short legs has lost the genetic information for longer legs.

This process of destroying or removing genetic information from the genome (total DNA) of the species has been used by plant breeders for many decades to create new varieties of plants with desirable new qualities.

## Impossible Genetics

Dr. Macki Giertyche, Professor Emeritus of Genetics, at Torun University, Poland, states:

> "The differing varieties within each species come from re-combination - from the mixing and concentrating of genes during sexual reproduction. This is not mutation, for they are simply drawing upon the gene pool already present in the species and concentrating different combinations of genes in the different varieties. Some people claim this to be an example of macro-evolution, and that through this ongoing process a new biological species can arise. This is not so! All that has happened is that some genes have been segregated out from the population, and that the resulting variety is impoverished - it is poorer in gene content. No new genes have been formed - and if there are no new genes, there is no potential for new organs; they are just a different variety of the same species...."[6]

Dr. Macki Giertyche continues:

> "Re-combination, or the mixing and concentrating of genes, does not provide new genes. FOR EVOLUTION TO OCCUR WE NEED NEW GENES, FULL OF NEW GENETIC INFORMATION. There is no natural process known to science which will produce new genes. It is impossible."[7]

The supposed "evolution" of varieties within a species is the result of genetic specialisation, brought about by the recombination of genes through sexual reproduction and the selection pressures upon the population as a result of environmental factors. This is simply variation and adaptation, not evolution, because it occurs **within** a species. The different breeds or varieties of dogs are all still dogs; members of the species *canis familiaris*. Similarly, the different varieties of mice are all still mice. Mice have not turned into dogs, and dogs have not transformed into elephants! The recombination and loss of genetic information has not led to the formation of a new species; just a new

## Impossible Genetics

variety *within* the same species. No new genetic information has been added to the DNA of the species.

Natural selection and artificial selection (selective breeding by people) are the main drivers of this process, but it is worth mentioning that the genetic diversity of a species can also be influenced by more subtle factors. Coming into contact with radiation, various chemical catalysts or biological stressors can alter the shape of the coiled DNA, resulting in genetic changes.[8] These catalysts, together with further gene recombination through sexual reproduction, can cause genes that were dormant to be switched on. Conversely, genes that were active can become dormant. Genes that were recessive can become dominant and genes that were dominant can become recessive.[9] The result is that certain traits are strengthened, while others are weakened, and if these changes are perpetuated and strengthened through further sexual reproduction, a new variety of the species may be formed. But, once again, no new genetic information has been added. It is still the same species.

It is possible for new genetic information to be transferred between varieties within a species. Plant breeders cross-pollinate in order to create a new variety within a species. This introduces the pollen from one variety to the stigma of another variety, thereby introducing new or missing genes to the plant being pollinated. Nature, however, has strong "anti-evolution" mechanisms in place to guard against cross-pollination of varieties that are too dissimilar. Within plants, a complex molecular recognition system ensures that pollen blown by the wind does not germinate on the stigma of a completely different species.[10] Only plants of very similar species can successfully cross-breed. Similarly, in the animal world, the ovum of one species will reject the sperm of completely different species. On rare occasions, such as the mating of two animals of a similar species or sub-species (such as horses and donkeys), the resulting offspring (mules) will be sterile and unable to perpetuate the new variety. In all of these cases of cross-fertilisation, however, no *new* genetic information has been *created*. Cross-fertilisation simply involves the transfer of **pre-existing** genetic information from one variety of organism to another.

Variation, therefore, is a naturally occurring phenomenon that allows a species to adapt to environmental pressures through genetic specialisation. This involves the loss of less desirable genes through genetic recombination in sexual reproduction, or the rendering of certain genes dormant through environmental pressures. It almost always involves the **loss** of genetic information and **never** involves the creation of completely **new** genetic information.

## GENETICS AND EVOLUTION

Charles Darwin's theory of evolution rests solely upon the premise that the processes involved in observable variation and adaptation can be extrapolated to explain the evolution of one species to a completely new species. On the basis of his observation of different beak sizes within finch populations (and other, similar observations), Darwin theorised that perhaps, given enough time, worms could turn into lizards and rats could turn into birds. According to his theory, fish eventually turned into elephants, giraffes and humans. To say that this is an extraordinary leap of logic is an understatement! Even Darwin, himself, recognised how tenuous and speculative his theory was:

> *"Therefore, I should <u>infer from analogy</u> that probably all organic beings which have ever lived on this earth have descended from one primordial form, into which life was first breathed."*[11]

Darwin freely admitted that his theory was not based upon solid evidence, but arose primarily from "analogy"; from his imaginative speculations:

> *"I am quite conscious that my speculations run quite beyond the bounds of true science."*[12]

## Impossible Genetics

So, was Charles Darwin correct? Despite his lack of hard evidence in 1859, did he stumble across the biological explanation for the development of species on our planet? Is the evolution of one species into a completely new species genetically possible?

The opinion of a growing number of the world's leading geneticists is a clear "NO".

In the opening chapter of this book, I briefly referred to the findings of the 1980 Chicago conference, and it is worth repeating here. In November 1980, at the Natural History Museum in Chicago, a large number of the world's leading geneticists held a seminar to consider whether the processes involved in variation and adaptation, (which they then termed "microevolution") could also produce the huge genetic changes required for the evolution of completely new species. The conference was demanded by many geneticists in response to the rapidly increasing understanding of the complexities of genetics and the new challenges that it posed for the theory of evolution. The findings of the conference were reported in the next issue of "Science" magazine, which stated;

> *"The central question of the Chicago conference was whether the mechanisms underlying microevolution can be extrapolated to explain the supposed phenomena of macroevolution. At the risk of doing violence to the opinions of some of the scientists at the meeting, the answer was a clear 'No'."*[13]

Since the 1980 Chicago Conference, our understanding of genetics has continued to grow, and so too has the chorus of voices expressing concern regarding the incompatibility of Darwin's theory with the science of genetics.

So, let us briefly examine the nature of that incompatibility.

**NEW GENETIC INFORMATION NEEDED**

In order for one species to evolve into a completely new species, a huge amount of **new** genetic information needs to be created and introduced to the genome (total DNA chain) of the existing species. Take, for example, the humble earthworm. For the earthworm to develop legs and eyes (of which the poor blind, belly-crawling creature currently has none), a vast number of new genes and genetic instructions need to be created. Dr. John Ashton, in his book, *"Evolution Impossible"*, comments:

> *"An example would be a worm evolving jointed legs so it could walk or developing eyes so it could see. These new features, when they first form, would require massive amounts of new genetic information to encode for all the parts of the legs, their control mechanisms, and the programming of the brain to use them. Similarly, with the first eye, all the components, the lens, focusing mechanisms, the optic nerve, the blood supply, and so on, would have to be encoded for in the DNA of the organism."*[14]

These new biological features - legs and eyes - could not possibly be created by merely shuffling the worm's existing genes. They require completely **new** genes. I can take my lawn mower apart and shuffle and reassemble the parts forever, but it will never turn into a motor vehicle or an aeroplane. For that to occur, a huge amount of **new** material and technology needs to be **added** to my lawn mower parts. The same is true of genetics.

The theory of evolution proposes that at some point in the distant past, some creatures grew legs, even though legs had never existed in the history of our planet prior to this. So, where did the genes for legs come from? Dr. Lee M. Spencer, in his book, *"Not by Chance! Shattering the Modern Theory of Evolution"*,[15] outlines the absolute impossibility of huge amounts of new genetic information coming into existence out of thin air.

# Impossible Genetics

It is worth considering how much new genetic information would need to be created for the addition of new biological components and the creation of more complex biological life. The following table* compares the amount of genetic information across a variety of biological lifeforms. Keep in mind, the genome size refers to the total number of base pairs within the organism's DNA, and the number of genes equates to the number of chapters within that book, with each chapter addressing a specific function or characteristic of the organism.

| COMPARISON OF GENETIC COMPLEXITY | | |
| --- | --- | --- |
| Organism | Genome Size (Base Pairs) | Genes |
| **E.Coli** | 4.6 million | 4,300 |
| **Roundworm** | 100 million | 20,000 |
| **Fruit Fly** | 140 million | 14,000 |
| **Mouse** | 2.8 billion | 20,000 |
| **Human** | 3.2 billion | 21,000 |

\* Based on information from the Genetics Science Learning Center, University of Utah.

Looking at this table, the single-celled bacterium, E.coli, has 4.6 million base pairs (letters) in its DNA chain, and 4,300 genes (clusters of "words", or "chapters"). Evolution teaches that life began as single-celled organisms such as this, and then evolved into increasingly complex lifeforms through natural selection and mutations. So let's have a look at a lifeform slightly higher up the supposed "tree of life" in terms of complexity; the tiny roundworm.

## Impossible Genetics

The roundworm has 100 million base pairs in its DNA. That is **21 times** more letters in its DNA, compared to E. Coli. For a single-celled organism such as E. Coli to eventually develop into a roundworm, it somehow needs to acquire **21 times more** coded information in its DNA. It requires an additional **95.4 million** base pairs in its genome (DNA)! And, as you can see from the table, it needs an additional **15,700 genes**; brand new sets of instructions for the formation of skin, digestive tract, muscles, blood, circulatory system, etc. Evolutionists have no plausible explanation for where this vast amount of new genetic encoding could have come from.

A fruit fly is more complex again, with 140 million base pairs. While it has less genes than the roundworm, its genes are much more complex (containing thousands more base pairs compared to the roundworm's genes - its "chapters" are much longer). It must also be understood that the fruit fly doesn't merely require an additional 40 million base pairs in its genome when compared to the roundworm, because the roundworm and the fruit fly don't share the roundworm's 100 million base pairs with an additional 40 million simply tacked on. A large percentage of the fruit fly's 140 million base pairs in its DNA are **completely different** from the base pairs of the E.coli bacterium (letters arranged in a very different order). This means a huge amount of additional, completely new genetic information must somehow be created; encoding for wings and legs and eyes and mouth and internal organs.

Compared to the fruit fly, a common house mouse has **20 times** as much genetic information, and humans have **23 times** as much.

The bottom line of all these numbers is this: Humans have **700 times** more genetic information in their DNA than a single-celled bacterium, an additional **3.195 billion** base pairs! If, as evolution proposes, humans ultimately evolved from single-celled life, where did this vast amount of additional genetic information come from? How did it get there?

# Impossible Genetics

Evolutionists are still scratching their heads over this issue. For example, the commonly used *"Biology"* textbook, by Solomon, Berg and Martin, states:

> *"One concern of macroevolution is to explain evolutionary novelties, which are large phenotypic changes."*[16]

In layman's terms, this says, *"We have no idea how new genes for new organs could have developed!"* Berkeley University website puts it more bluntly:

> *"Biologists are trying to figure out how evolution happens."*[17]

Of course they are. Because it **didn't** happen!

In July 2008, a conference of the world's leading geneticists was held to discuss the problem of how natural processes could produce the vast amounts of new genetic coding necessary for evolution to take place. The conference, in Altenburg, Austria, could find no viable natural means by which this new information could be created. After the conference, Dr. Jerry Fodor, of Rutgers University, is quoted as saying, *"Basically I don't think anybody knows how evolution works."*[18]

Even outspoken atheist, Dr. Richard Dawkins, has been unable to offer a viable explanation for how vast amounts of new genetic information could be added to the gene pool via natural means. Dr. Alex Williams, in his paper "Inheritance of Biological Information Part II: Redefining the Information Challenge"[19] refers to Dr. Dawkins' inability to provide an adequate explanation:

> *"When asked by creationists if he knew of any biological process that could increase the information content of a genome, Oxford Professor Richard Dawkins could not answer the question.[20] He later wrote an essay of the subject titles, "The Information Challenge"*[21]*, but even in the essay he could not give a single*

*example of a mutation that could increase the information content of a genome."*

The important point to glean from all of this, is really quite simple. For evolution to work, huge amounts of completely new genetic information must be created, not just for the evolution of humans, but for the evolution of every one of the millions of species on our planet! And, by *"new genetic information"*, I mean genetic information that had previously never existed, for organs and biological functions that have never previously existed.

Furthermore, scientists do not observe new genes for new organs or functions being created today, nor can they cite even one clear example of speciation (one species turning into a completely new species with new physical features) having been observed in the past.

Significantly, Berkeley University's *"Understanding Evolution"* website has a whole page dedicated to *"Examples of Microevolution"* immediately following their page, *"What is Microevolution?"*. But after their page, *"What is Macroevolution?"* there are no examples of macroevolution (one species evolving into a completely new species) offered at all. This is not surprising, as there have **never** been **any** observed or recorded instances of macroevolution.[22]

The Wikipedia page on macroevolution suggests only two examples; the breeding of horses with donkeys to produce sterile mules, and the ability of viruses to transfer genetic material from one species to another. Yet both these examples are not producing ANY new genetic material; they are simply recombining existing genetic material across species. Neither of these are examples of macroevolution, which requires huge amounts of *new* genetic material, *which has never existed before,* to be created.

Similarly, in Dr. Richard Dawkins' 470-page book, *"The Greatest Show on Earth: The Evidence for Evolution"*, he cites only one possible example of macroevolution.[23] He quotes experiments on E. coli bacteria by Dr

Richard Lenski, where over 31,000 generations of the bacteria were observed, resulting in a mutation enabling it to utilise citrate. However, the conclusion of Lenski himself is that the mutation simply involved a dormant gene in the bacteria being activated, thus no new genes were actually produced.[24] Furthermore, the E. coli was still E. coli. It had not become something different. So, the only example of macroevolution that the world's leading proponent of evolution could produce, turns out to be just another example of adaptation and variation!

Not only are scientists unable to produce a single example of macroevolution, but, to date, they have been unable to propose a reasonable process by which it could have occurred in the past. The current great hope of evolutionists in this regard, is the idea that genetic mutations might be able to create new genetic information. However, that hope is fading fast, as the impossibility of mutations producing incremental beneficial changes becomes increasingly clear.

**THE IMPOSSIBILITY OF POSITIVE GENETIC MUTATIONS**

Evolutionists theorise that new genes and genetic information may be able to be created through random mutations of existing genetic material. There are several major flaws in this theory, which we will now explore in detail:

1. Genetic repair mechanisms naturally correct and delete mutations

2. Mutations are almost never beneficial

3. Impossible amounts of new material are needed

4. Incremental mutations would not be retained by the organism

5. Evidence of rapid genetic entropy

6. Mutations are not genetic script writers

**1. Genetic Repair Mechanisms**

There are inbuilt genetic repair mechanisms within all cells, and within the genome (DNA) itself, which naturally correct or delete mutations. All biological life has a pre-programmed ability to protect the organism against mutations.

A genetic mutation is when a section of the DNA is altered, either through damage caused by DNA interrupters, such as ultraviolet radiation and mutagenic chemicals (chemicals that cause mutations), or by replication errors when the DNA is copying itself during cell division.[25] Genetic mutations can be as small as a single base pair in the DNA or can involve changes to whole segments, encompassing several genes.

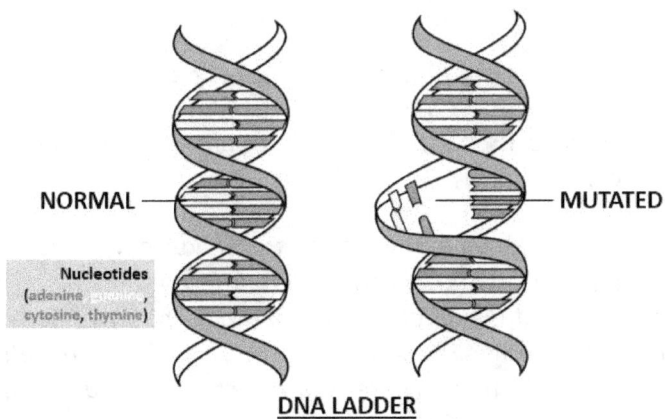

Diagram showing DNA mutation, sourced from James Madison University website.

Mutations in the DNA occur constantly and are almost always corrected. The 3.2 billion base pairs that comprise our double-helix DNA are almost never perfectly copied into a new cell during cell division and reproduction. Some scientists estimate that a newly divided single cell can initially contain up to 120,000 replication errors that need to be corrected by the cell by means of an initial repair phase known as "proofreading", and a secondary correction phase referred to as "mismatch repair".[26] During this two-phase repair process, polymerase enzymes (tiny molecular machines) act upon a new cell's DNA, realigning mismatched nucleotides (sections of DNA) and adjusting incorrectly copied DNA segments until the whole double helix is an exact replica of the original.[27] But this process is not perfect. Some mistakes do "slip through the net" and are perpetuated in subsequent generations of the cell. The precise location of these unfixed errors within the 3.2 billion base pairs of the DNA strand, as well as the extent of the errors (how many base pairs have been damaged), will determine whether the gene that may be affected will have limited function or no functionality at all.

This brings us to the main flaw in the evolutionists' argument regarding mutations:

**2. Genetic Mutations Are Almost Never Beneficial**

Dr. Macki Giertyche, Professor Emeritus of Genetics, at Torun University, Poland, states:

> *"Evolution is not a science; it is a philosophy... The science of genetics shows that macro-evolution is not possible... The main argument of evolutionists is that small, positive (or beneficial) mutations occur in the reproduction cells and are retained by natural selection. These mutations are said to gradually accumulate over time until a new species is formed. <u>Now, I am a geneticist, and I can confirm that in all the studies in all the laboratories around the world, where many generations of organisms have been observed, nowhere have positive mutations ever been observed. All mutations are either neutral or harmful - they are never an improvement.</u>"*

This is a significant problem for evolutionists. For example, missing sections of DNA and errors in the encoding and duplication of genes is the cause of a wide range of diseases and syndromes, including cystic fibrosis, Huntington's disease, haemophilia, Down syndrome and familial high cholesterol.[28] Genetic mutations do not create, they corrupt! Overwhelmingly, genetic mutations are "going the wrong way". Almost every mutation creates disease, dysfunction or abnormality. As biologist Dr. Gary Parker comments, in his article, *"Mutations - Yes; Evolution - No"*:

> *"Mutations are harmful at least 1000 times more often than they are helpful. No evolutionist believes that standing in front of x-ray machines would eventually improve human beings. No evolutionist argues that destruction of the earth's ozone layer is good because it increases mutation rates and, therefore, speeds up evolution. Evolutionists know that decreases in the ozone layer will increase mutation rates, but they, like everyone else, recognize that this will lead only to increased skin cancer and to other harmful changes. Perhaps a helpful change might occur [occasionally], but it would be drowned in the sea of harmful changes."*[29]

The most common claim by evolutionists of positive genetic mutations, revolves around mutations observed in viruses and bacteria. Antibiotic resistance that develops within bacteria is often quoted as an example of a beneficial mutation (beneficial to the bacteria!). But Dr. Gary Parker explains that this does not occur as a result of mutation, but is simply a result of recombination and activation of genes that are already present within the bacteria.[30] No new genetic information has been created that was not already present in some form.

### 3. Impossible Amounts of New Material Are Needed

For evolution to be possible, tens of thousands of new genes, containing thousands or millions of base pairs of information, need to be created by random chance. For this to occur as a result of genetic mutation, it would require trillions and trillions of generations of DNA replication, producing tens of thousands of incremental mutations, in strict sequence, in order to produce a whole, fully functioning new gene which was coded for a new organ or biological function. Each mutation would have to be retained by the organism then added to, in precisely the right sequence, mutation by mutation, with each new mutation being retained in precisely the right location within the DNA, without it being repaired or deleted by the natural repair mechanisms, until a whole new gene was created.

Given the random nature of mutations that "slip through" the repair process, the mathematical probability of a whole new gene being created sequentially is such that it is completely impossible. Dr Gary Parker explains:

> *"The mathematical problem for evolution comes when you want a series of related mutations. The odds of getting two mutations that are related to one another is the product of their separate probabilities: one in 10 to the power of 7 x 10 to the power of 7, which equals 10 to the power of 14. That's a one followed by 14 zeroes, or 100 trillion! Any two mutations might produce no more than a fly with a wavy edge on a bent wing. That's a long way from producing a truly new structure, and certainly a long way from changing a fly into some new kind of organism. You need more mutations for that. So, what are the odds of getting three mutations in a row? That's one in a billion trillion (10 to the power of 21). Suddenly, the ocean isn't big enough to hold enough bacteria to make it likely for you to find a bacterium with three simultaneous or sequential related mutations.*
>
> *What about trying for four related mutations? One in 10 to the power of 28. Suddenly, the earth isn't big enough to hold enough organisms to make that very likely, and we're talking about only*

*four mutations. It would take many more than that to change a fish into a philosopher, or even a fish into a frog."*[31]

What about the creation of a complex animal by random mutations? It has been estimated that the creation of an animal such as the horse, from the humble beginnings of a single-celled organism, (which evolution claims is the common ancestor of all lifeforms), would require 10 to the power of 3 million generations of DNA replications, in order for the necessary random mutations to occur.[32] We don't even have a name to give to this number, it is so big! And even then, all the mutations would have to occur in exactly the right sequence. But if the earth is 4.5 billion years old, as evolutionists maintain, that equates to only 10 to the power of 17 seconds. If 10 to the power of 3 million generations of DNA replications are required for a complex animal to arise by chance, but the universe has only existed for 10 to the 17 seconds, it would require over 10 to the power of 2 million DNA replications **every second** in order for the necessary number of DNA replications to have occurred. To put it another way, the earth would need to have existed untold quadrillions of times longer than it supposedly has, if random mutations can come close to explaining the development of species!

These huge mathematical impossibilities are widely recognised as being a serious problem for the theory of evolution, even among evolutionists. For example, Drs. Paul S. Moorehead and Martin M. Kaplan, in their published work, *"Mathematical Challenges to the Neo-Darwinian Interpretation of Evolution"*,[33] discuss this issue as a major problem for evolution.

### 4. Incremental Mutations Would Not Be Retained By The Organism

Even if genetic mutations could somehow incrementally produce new genetic information, the intermediate incremental stages of new genes would weaken rather than strengthen the organism. In fact, a half-formed gene would have no viable, beneficial function until the gene was completed in full. Until then, it would need to be retained by the organism for trillions of replications, without being corrected or deleted by the natural repair mechanisms within DNA. Furthermore, in its partially formed state, it would almost certainly be deleterious (harmful) to the organism. As the number of incremental mutations increased in the supposedly new, yet inoperable, gene, the possible harmful effects would multiply, to the point where the cells containing those gross accumulated errors would almost certainly cease to be viable.

Furthermore, on the biological level, intermediate species resulting from supposed evolutionary mutation would simply not survive. For example, a reptile which has somehow (miraculously!) managed to create dozens or even hundreds of new genes for all the biological elements of wings (new bone structures, new blood vessels, new muscles, new skin, new feathers etc), would be at an extreme disadvantage during all but the final stage of its transformation. With front legs or arms that are half-way evolved to wings, it would not yet be able to fly and would no longer be able to function and survive effectively with increasingly malformed front legs or arms. Such a creature would be weaker, not stronger, and would likely perish through natural selection.

PROF. GUISEPPE SERMONTE, an Italian biochemist, geneticist, and molecular biologist states:

> "Recent discoveries in molecular biology have deeply undermined the theory of evolution. The claim of evolution - that mutations are retained and strengthened by natural selection - is not true. What natural selection does is eliminate genetic mutations..."[34]

### 5. Evidence of Rapid Genetic Entropy

The overwhelming tendency of genetic mutations to produce harmful rather than beneficial changes is obvious when the human genome (DNA) of modern humans is compared with that of people in the past. When DNA samples are studied from mummified remains and bone tissue samples from people living hundreds and even thousands of years ago, and compared to modern DNA, we find that our genome today contains significantly more harmful genetic mutations.

The accumulation of genetic defects within the human genome, resulting from harmful mutations in our DNA, is observable from generation to generation. Biologists estimate that between 100 and 300 genetic defects enter the human genome (the 3.2 billion base pairs) every generation (about 20 years).[35] Individual base pairs and sections of the DNA are not copied correctly during sexual reproduction and cell division, and errors accumulate in the DNA. As most of these are not in areas of the genome that encode for proteins, they are effectively invisible to the processes of natural selection that would otherwise delete them, so they accumulate within the genome of a population from generation to generation.

In other words each of us have 100 to 300 more wrongly encoded sections of DNA than our parents did, and they had 100 to 300 more defects than their parents, and so on.

Not all errors in the DNA are immediately serious. Many occur in parts of the genome for which geneticists cannot currently find any obvious function. These segments of the genome are referred to as "non-coding DNA" (and sometimes colloquially called "junk DNA"). However, the rapidly evolving field of epigenetics is revealing that even these non-coding segments of DNA can become activated as a result of environmental factors, and so errors within these segments can then have a deleterious effect. Other errors are more immediately harmful, occurring within active segments of the DNA, and negatively impacting health, as we have already noted. This is almost certainly why we see increasing intolerances and disorders among children today compared to even a few generations ago.

This is also evident in the decreasing lifespans of biblical patriarchs after the flood. Although the original humans were created with a perfect genome, the accumulation of mutational errors led to a rapid decline in longevity:

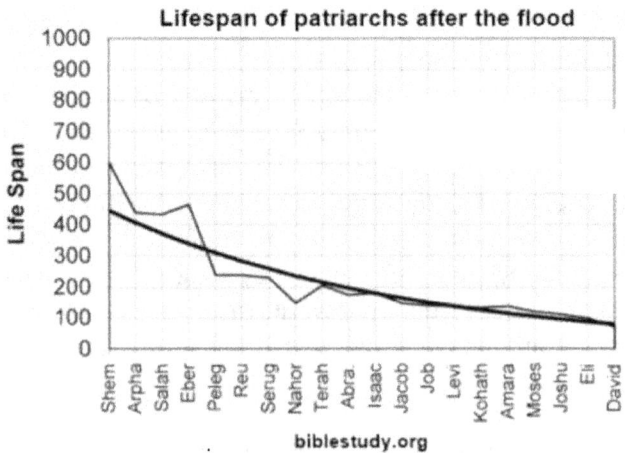

The fact that humans live longer today, on average, than they did 100 years ago, is a result of improved diet and hygiene, as well as advances in medicine in fighting diseases, all of which have compensated, to some extent, for our gradual decline in genetic health.[36]

In his book, *"Genetic Entropy and the Mystery of the Genome"*, Dr. John Sanford provides a devasting critique of evolution by pointing out that genetic mutations, rather than resulting in beneficial "upward" evolution, overwhelmingly result in "downward" evolution, or "devolution". He points out that the inter-generational, observable, rapid accumulation of deleterious mutations absolutely sweeps away any remote possibility of beneficial mutations being able to produce higher, more complex lifeforms.

Furthermore, Dr Sanford and others, such as Dr. Jeffrey P. Tomkins,[37] point out that the rapid accumulation of genetic defects within the human population is further evidence of a young earth. Whereas evolutionists claim that early humans such as homo erectus appeared about 2 million years ago, and "modern" humans (homo sapiens) appeared about 300,000 years ago[38], Sanford and Tomkins maintain that humans simply could not have existed on earth for hundreds of thousands of years, because we would have become extinct well before that. Dr. Alex Williams, in his paper, *"Mutations: Evolution's Engine Becomes Evolution's End"*, estimates that, given our current rate of genetic defect accumulation, an initial population of 100 million would become extinct in 32,800 years.[39] As the same process of accumulated genetic deterioration is also at work within all animals[40], this means that biological life on earth cannot be millions of years old.

**6. Mutations Are Not Genetic Script Writers.**

Another major problem facing the mutation theory of evolutionary development, is the simple fact that a genome (DNA) has to exist in the first place in order for mutation to occur. Mutations typically damage one or two points in already existing DNA molecule. But mutations cannot bring about the existence of the DNA in the first place. Mutations damage the script of a gene; they do not write it.

Dr. John Sanford explains:

> *"Mutations are NOT genetic "script writers"; they are merely "typographic errors" in a genetic script that has <u>already been written</u>. Typically, a mutation changes only one letter in a genetic sentence averaging 1,500 letters long."*[41]

Mutations cannot write whole sections of brand-new genetic script (instructions), creating whole new genes. Nothing like this has ever been observed to happen and no known process could make it even a remote possibility.

As Dr Sanford goes on to say:

> *"To make evolution happen—or even to make evolution a theory fit for scientific discussion—evolutionists desperately need some kind of "genetic script writer" to increase the quantity and quality*

*of genetic INFORMATION. <u>Mutations have no ability to compose genetic sentences, no ability to produce genetic information, and, hence, no ability to make evolution happen at all.</u>"*[42]

## CONCLUSION

The science of genetics is leading a growing number of scientists to abandon evolution as a viable theory for explaining life on earth. These include scientists who are not Christians. For example, Dr. Chandra Wickramasinghe, professor of Applied Mathematics and Astronomy, at University College, Cardiff, UK, is a Buddhist and was once a convinced evolutionist. While he still does not ascribe to the Christian creationist viewpoint, he has come to completely reject evolution as a viable theory. Famously, he appeared as an expert witness in a court case in Arkansas, USA, in 1981, which was attempting to determine whether the theory of evolution should be the only theory of origins taught in state schools. He offered damning scientific testimony against evolution, stating:

> *"The reason for my standing here to challenge accepted beliefs in Darwinian evolution ... successive copying would accumulate errors, but such errors on the average would lead to a steady degradation of information. It is ridiculous to suppose that the information provided by one single primitive bacterium can be upgraded by copying to produce a man, and all other living things that inhabit our planet. This conventional wisdom, as it is called, is similar to the proposition that the first page of Genesis copied billion upon billions of time would eventually accumulate enough copying errors and hence enough variety to produce not merely the entire Bible but all the holding of all the major libraries of the world. The two statements are equally ridiculous."*[43]

The recent science of genetics has unveiled the vast and unimaginably complex amount of genetic data required for the formation of even the smallest of organs, and, for anyone with an open mind, reveals the stark impossibility of life originating and evolving by random processes.

There is a fascinating admission in the *"Introduction"* to the 6th edition (1956) of Darwin's *"On The Origin of Species"*, written by Dr. W.R. Thompson, Entomologist and Director of the Commonwealth Institute of Biological Control, Ottawa, Canada. He wrote;

> *"Darwin himself considered that the idea of evolution is unsatisfactory unless its mechanism can be explained. I agree, <u>but since no one has explained to my satisfaction how evolution could happen I do not feel compelled to say that it has happened</u>. I prefer to say that, on this matter, our investigation is inadequate."*[44]

In other words, even a scientist who is positively *inclined* towards Darwin, to the point where he was invited to write the introduction to the 6th Edition, could not bring himself to say that the theory is, after all this time, supported by hard evidence! Because there is none! Hence, his frank admission of doubt; *"I do not feel compelled to say that it [evolution] has happened"*.

The science of genetics shows the impossibility of evolution, and points to the work of an intelligent Creator.

---

**ENDNOTES - Chapter 8**

---

[1] Ken Ham, "The Evolution Tapes", Creation Science Foundation, 1980.

[2] John F. Ashton. "Evolution Impossible", Green Forest, AR, Masterbooks, 2013, p.51

[3] John F. Ashton. "Evolution Impossible", Green Forest, AR, Masterbooks, 2013, p.51

[4] Waterland R.A., Jirtle R.L., Transposable elements: targets for early nutritional effects on epigenetic gene regulation, *Mol Cell Biol.* **23**(15):5293-300, 2003. Cited in https://creation.com/epigenetics-challenges-neo-darwinism

[5] Waterland R.A., Jirtle R.L., Transposable elements: targets for early nutritional effects on epigenetic gene regulation, *Mol Cell Biol*. **23**(15):5293-300, 2003. Cited in https://creation.com/epigenetics-challenges-neo-darwinism

[6] "Unlocking the Mystery of Life" DVD, Illustra Media, 2010

[7] IBID

[8] John F. Ashton. "Evolution Impossible", Green Forest, AR, Masterbooks, 2013, p.53.

[9] "Dominant Vs Recessive Genes", https://genetics.thetech.org/ask/ask45, 2nd Feb 2012,

[10] John F. Ashton, "Evolution Impossible", Master Books, Green Forest, AR, 2013, p.55

[11] Charles Darwin, "The Origin of Species", London, John Murray, 1859, p.156

[12] From a letter to Asa Gray, Harvard biology professor, cited in "Charles Darwin and the Problem of Creation", N.C. Gillespie, p.2

[13] Roger Lewin, "Science" Journal, Vol. 210(4472), 1980, pp.883-887

[14] John F. Ashton. "Evolution Impossible", Green Forest, AR, Masterbooks, 2013, p.56.

[15] Lee M. Spencer, "Not by Chance! Shattering the Modern Theory of Evolution", New York, Judaica Press, 1997.

[16] Eldra P. Solomon, Charles E. Martin, Diana W. Martin, Linda R. berg, "Biology", Stamforn, CT, cengage learning, 2015, 10th Edition, p.130

[17] Evolution 101, 2009, "The Big Issues", evolution.berkeley.edu/evosite/evo101/VIIBigissues.shtml

[18] Susan Mazur, The Altenberg 16: "An Expose of the Evolutionary Industry", Berkeley, CA, North Atlantic Books, 2010.

[19] Alex Williams, "Inheritance of Biological Information Part II: Redefining the Information Challenge", https://creation.com/images/pdfs/tj/j19_2/j19_2_36-41.pdf

[20] Richard Dawkins interviewed in the video From a Frog to a Prince, video produced by Keziah, available at

[21] Dawkins, R, The 'Information Challenge', The Skeptic 18(4), 1998; , 3 June 2005

[22] https://evolution.berkeley.edu/evolibrary/article/0_0_0/evoscales_03

[23] Richard Dawkins, "The Greatest Show on Earth: Evidence for Evolution", New york, NY, Simon & Schuster, 2010, p.131

[24] Quoted in John F. Ashton, "Evolution Impossible", Green Forest, AR, Masterbooks, 2013, p.57.

[25] https://ghr.nlm.nih.gov/primer/mutationsanddisorders/genemutation

[26] https://www.khanacademy.org/science/high-school-biology/hs-molecular-genetics/hs-discovery-and-structure-of-dna/a/dna-proofreading-and-repair

[27] https://www.khanacademy.org/science/high-school-biology/hs-molecular-genetics/hs-discovery-and-structure-of-dna/a/dna-proofreading-and-repair

[28] John F. Ashton, "Evolution Impossible", Green Forest, AR, Masterbooks, 2013, p.61.

[29] Gary Parker, "Mutations -Yes; Evolution - No", https://answersingenesis.org/genetics/mutations/mutations-yes-evolution-no/

[30] Gary Parker, "Mutations -Yes; Evolution - No", https://answersingenesis.org/genetics/mutations/mutations-yes-evolution-no/

[31] https://answersingenesis.org/genetics/mutations/mutations-yes-evolution-no/

[32] Gary Parker, "Mutations -Yes; Evolution - No", https://answersingenesis.org/genetics/mutations/mutations-yes-evolution-no/

[33] Paul S. Moorehead and Martin M. Kaplan, Mathematical Challenges to the Neo-Darwinian Interpretation of Evolution, Wistar Symposium No. 5 (Philadelphia, PA: Wistar Institute Press, 1967)

[34] "Unlocking the Mystery of Life" DVD, Illustra Media, 2010

[35] http://evolutiondismantled.com/meltdown

[36] https://creation.com/genetic-entropy

[37] https://www.icr.org/article/genetic-entropy-points-young-creation

[38] https://en.m.wikipedia.org/wiki/Homo

[39] Alex Williams, "Mutations: Evolution's Engine Becomes Evolution's End", https://creation.com/mutations-are-evolutions-end

[40] Alex Williams, "Mutations: Evolution's Engine Becomes Evolution's End", https://creation.com/mutations-are-evolutions-end

[41] https://answersingenesis.org/genetics/mutations/mutations-yes-evolution-no/

[42] https://answersingenesis.org/genetics/mutations/mutations-yes-evolution-no/

[43] Chandra Wickramasinghe, Testimony in court case, Arkansas, USA, 1981, quoted in https://www.panspermia.org/chandra.htm

[44] Introduction to "The Origin of Species", 6th Edition (1956) p. 25

Chapter 9

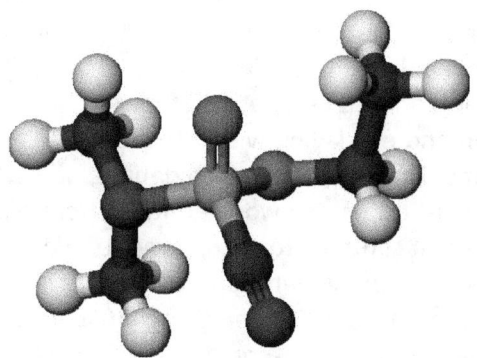

# The Origin Of Life

Without a doubt, the most difficult problem facing evolutionists, or anyone who does not believe in a Creator God, is explaining how non-living matter could have given rise to organic life in the very beginning. Evolution teaches that the vast array of species in existence today ultimately originated from single-celled organisms which somehow began to exist. Yet that "somehow" is precisely the problem. How did the first living cell form? What possible process could account for the creation of a living biological organism from dead, inanimate matter? To the present day, scientists have not been able to propose a viable means whereby this could have occurred, nor have they been able to create anything remotely close to a living cell in laboratories, even with the most advanced technology available to us today.

# The Origin of Life

**SPONTANEOUS GENERATION**

In the Middle Ages, many scientists and philosophers believed in the spontaneous generation of life. Maggots supposedly arose from decaying meat, frogs from stagnant ponds, earthworms from manure, insects from the morning dew and mice from warm, moist soil. This was the prevailing view from the time of Aristotle until the mid-19th century, when various experiments provided strong evidence that what had previously been interpreted as spontaneous generation of life had simply been the result of contamination and egg-laying within those various mediums by pre-existing organisms. There was still much debate, however, and, in 1859, Dr. F. Pouchet published a major work, entitled *"Heterogenie"*,[1] providing answers to the recent contradictory experiments and arguing strongly for the validity of spontaneous generation. Thus, it was in the same year that Charles Darwin published his *"On the Origin of Species"*, at a time when there was still significant, widespread belief that life could arise from non-living matter by chance natural processes. Darwin had grown up in a world that believed in spontaneous generation, and it is this belief in self-perpetuating natural causes that underpins his theory. It was only three years later, in 1862, that Louis Pasteur published the findings of his brilliant set of experiments, finally putting an end to the superstitious belief of spontaneous generation. Unfortunately, by then, Darwin's theory of natural selection had taken off and gathered an unstoppable momentum of its own.

If spontaneous generation of life from non-living matter does not occur, however, this leaves the atheist with a problem. How can the creation of life in the very beginning be explained? Dr. George Wald (1906 - 1997), a biologist who won the Nobel Prize in Physiology in 1967, commented on this quandary:

> *"We tell this story [the overturning of belief in spontaneous generation] to beginning students of biology as though it represents a triumph of reason over mysticism. In fact it is very nearly the opposite. The reasonable view was to believe in*

*spontaneous generation; the only alternative, to believe in a single, primary act of supernatural creation. There is no third position. For this reason many scientists a century ago chose to regard the belief in spontaneous generation as a 'philosophical necessity.' It is a symptom of the philosophical poverty of our time that this necessity is no longer appreciated. Most modern biologists, having reviewed with satisfaction the downfall of the spontaneous generation hypothesis, yet unwilling to accept the alternative belief in special creation, <u>are left with nothing</u>."[2]*

Or are they left with nothing? Forced to reject the creationist view of the origin of life, modern scientists have reinvented the concept of spontaneous generation under a new name: abiogenesis.

## ABIOGENESIS

Abiogenesis refers to the supposed process whereby non-living, inorganic matter gave rise to the first living organisms in the beginning of earth's biological history. Russian biochemist, Dr. Alexander Ivanovich Oparin, was one of the strong proponents of the abiogenesis theory in the 20th century. He proposed that the chemical composition of primordial earth's atmosphere and oceans was such that everything needed for the construction of the simplest living cell was present. He theorised that, given sufficient time, and with the addition of heat and electrical energy from lightning, chemicals could combine and form simple molecules, and, with further time, these molecules could arrange themselves into increasingly complex amino acids and other building blocks necessary for the creation of the first cell. Oparin's landmark book, "The Origin of Life",[3] published in 1967, outlined his theory of the possible chemical evolution of life, building upon the theories and experiments of several other scientists in the preceding decades. In his book, he comments:

> *"A careful survey of experimental evidence reveals that it tells us nothing about the impossibility of generation of life at some other epoch or under some other conditions."*[4]

Dr. Stanley Miller's experiments in the mid-1950s, initially generated great excitement. He heated and electrically sparked a mixture of methane, ammonia, hydrogen and water, and managed to produce some very basic amino acids.[5] However, subsequent experiments completely failed to show how all the more complex chemical components of a cell could spontaneously form, let alone how they could then all arrange themselves into the precisely balanced, steady-state equilibrium necessary for a cell to form and become a living organism. To use an analogy, Miller discovered how to make a few small rocks, but failed to demonstrate how they could turn themselves into a cathedral.

To this very present day, scientists are unable to theorise or replicate how a simple cell could arise from non-living matter by chance. Significantly, the Wikipedia page for Abiogenesis states:

> *"Abiogenesis, or informally the origin of life, is the natural process by which life has arisen from non-living matter, such as simple organic compounds. While the details of this process are still unknown, the prevailing scientific hypothesis is that the transition from non-living to living entities was not a single event, but a gradual process of increasing complexity that involved molecular self-replication, self-assembly, autocatalysis, and the emergence of cell membranes. Although the occurrence of abiogenesis is uncontroversial among scientists, there is no single, generally accepted model for the origin of life, ... for how abiogenesis could have occurred."*[6]

This comment on Wikipedia is typical of the unfounded evolutionary belief in abiogenesis. Evolutionists are effectively saying, *"we don't know <u>how</u> it happened, we haven't <u>witnessed</u> it happening today, there is no <u>evidence</u> that it ever happened in the past, and we haven't been able to* _replicate_ it in our laboratories, but it _must_ have happened, because the only alternative is to believe in a Creator!"

Despite the failure of scientists to either propose or demonstrate how abiogenesis could possibly have occurred, unsupported statements claiming abiogenesis to be a "fact" abound in textbooks and academic papers. For example, in an article in the Scientific American journal, Dr. Jack Szostak, genetics professor at Harvard University, wrote:

> "... the first life arose from non-living matter around 3.7 billion years ago."[7]

Similarly, the popular university Biology textbook by McKee and McKee states:

> "The earth was formed from a cloud of condensing cosmic dust and gas about 4.5 million years ago. Life arose soon thereafter."[8]

But could life really have arisen on its own? Could *"dust and gas"* ultimately give birth to complex living organisms?

**THE AMAZING CELL**

The jaw-dropping complexity of the single living cell renders any notion of its creation via random processes utterly ridiculous. No amount of stirring of the primordial soup and no number of lightning strikes could ever transform dead chemicals into a living organism.

In a previous chapter we investigated the concept of irreducible complexity; that when organic life is reduced to its simplest level, the single cell, it retains an astonishing complexity of interdependent, irreducibly complex components that all have to exist simultaneously and function harmoniously for the cell to be alive:

**DNA**

The complex genetic encoding inside every cell, containing billions of specific biological instructions in a precise order, could never be created by the random mixing of chemicals. The book you hold in your hands at the moment only contains about 80,000 words. No sane person would believe that the book formed itself by chance: that wood pulp formed itself into pages and glued themselves together, and then ink coalesced in mid-air and fell randomly onto the pages, forming letters, words, sentences, paragraphs and chapters - all in the precisely required order. (Similarly, if you are reading an eBook version, the infrastructure required to render the 80,000 words and images to your screen in a meaningful way required significant intelligent design). The DNA inside every cell in your body contains **40,000 times** more information than this book! It contains the entire blueprint of your body, together with instructions for its proper functioning, care and repair. As a piece of encoding, it is millions of times more complex than any computer program that has ever been written, and it testifies to the existence of an intelligent designer far superior to us. It is complete nonsense to speak of DNA coming into existence by chance.

**Molecular Machines**

As discussed in a previous chapter, each cell is a veritable biological factory, swarming with microscopic molecular machines constantly carrying out their pre-programmed functions in order to make the cell a living organism. This all takes place at the molecular level: hundreds of molecular machines of different shapes and sizes all working together.

Dr. George Javor, professor of biochemistry at Loma Linda University, California, has calculated the following statistics regarding the number and type of molecular components inside a single E. coli cell:[9]

- 2.4 million protein molecules made up of 4,000 **different** types of proteins

# The Origin of Life

- 255,000 nucleic acid molecules made up of 660 different types of nucleic acids

- 1.4 million polysaccharide molecules (a type of sugar) made up of 3 different types

- 22 million lipid (fat) molecules made up of 100 different types

- As yet uncounted millions of metabolic intermediate molecules made up of 800 different types

- As yet uncounted millions of mineral molecules made up of 30 different types

So, a single E. coli cell has 26 million functioning molecules, comprised of approximately 4,750 **different types** of molecules.[10] And each of those distinct types of molecules is comprised of smaller biopolymer building blocks, arranged in a precise order and pattern so that the molecule has the correct shape to carry out its function within the cell. A single protein molecule, for example, is comprised of up to 300 amino acids, arranged in very precise order.[11] Furthermore, in order for a cell to be alive, this army of precisely engineered molecular machines are constantly carrying out approximately 800 simultaneous chemical reactions.[12]

The problem facing evolutionists, is to explain how all this could come into existence by chance, from dead matter. Dr. John F. Ashton expresses the problem clearly:

> *"For the first life to start from non-living matter, thousands of specialised large complex molecules must somehow be synthesised in very large numbers from simple, small inorganic molecules. These molecules then have to come together randomly, over and over again, until somehow the structure of the cell is formed. This remarkable and complex structure would still, however, not be alive. To become alive, hundreds of*

> metabolic reactions would have to be initiated, with the metabolic intermediates already in place at just the right concentrations so that the reactions went the right way. Common sense tells us that these sorts of reactions don't just happen by chance - in fact, we cannot even make them happen."[13]

## MATHEMATICAL IMPOSSIBILITY

The fall-back position of evolutionists is that, given enough time, random processes must eventually result in the right chemical combinations to bring about life. Dr. Clifford D. Sirnack, in his book, *"Trilobites, Dinosaurs and Man"*, exemplifies this argument:

> "If we had amino acids, we then would have proteins, and if we had proteins, we would be well along the road to life. Given trillions upon trillions of possibilities for chemical combinations, given a few million years for it all to happen, the components of life would have appeared. And once that had been accomplished, once the bricks and the stones and the lumber for the building of life were present, then all that would have been required were a few more million years for life to actually appear."[14]

This type of argument appears, on the surface, to have merit. After all, surely given enough time, all the necessary components of cellular life can form by chance.

However, this line of reasoning vastly underestimates the statistical improbability of even a single protein evolving by chance. In 1962, Dr. Isaac Asimov (a professed atheist) calculated the probability of a simple insulin-like protein evolving by random chemical combinations.[15] He calculated that there are 8 x 10 to the power of 27 (8 followed by 27 zeros) different possible combinations of the building blocks that comprise a simple protein. In other words, there are 8 x 10 to the power of 27 possible combinations for these building blocks being put together, but only ONE of those combinations would create a viable protein. Even

## The Origin of Life

if we assume that a new combination has formed every minute that the universe has existed, after 14 billion years (the current evolutionist estimate of the age of the universe) only 4 x 10 to the power of 17 possible combinations would have formed.[16] The universe would have to be 10 BILLION TIMES OLDER than it supposedly is, in order to produce a single, simple protein!

In the case of haemoglobin, the probability of its chance formation is even more absurd. Dr. Asimov calculated that the possible chemical combinations for the formation of a single haemoglobin cell are 135 to the power of 165 (135 followed by 165 zeros).[17] Let us assume a ridiculously high reaction rate: let us say that 10 to the power of 100 (10 followed by 100 zeros) different combinations of the basic building blocks are produced EVERY SECOND. (I told you it was ridiculously high!). Even at this rate it would take ten trillion trillion years to produce all the necessary combinations to ensure the creation of the right one. That is TRILLIONS of times older than our universe supposedly is.

When it comes to something as complicated as the long DNA molecule, the probability of it forming by chance becomes even more ridiculous. Let us not even bother with human DNA, which contains 3.2 billion base pairs (pairs of chemicals joined together and arranged in a spiralling double helix). The simplest DNA in the smallest known virus contains only 5,000 base pairs.[18] The possible combinations for this, simplest of DNA molecules, is 10 to the power of 1,505 (10 followed by 1,505 zeros)![19] Given that the universe has supposedly existed for only 4 x 10 to the power of 17 seconds (4 followed by 17 zeros),[20] the formation of DNA by chance is absurdly impossible.

In dealing with statistical probabilities, we must also realise that there is a limit beyond which an event may be assigned a probability in theory, but be considered impossible in practice. Take for example, tossing a coin. The chance of tossing heads is 1 chance in 2. The chance of tossing two heads in a row is 1 chance in 4. The chance of 10 in a row is 1 in 1,024. The chance of 100 in a row is 1 in

1,268,000,000,000,000,000,000,000,000,000 (there isn't even a name for this number!). Although the mathematics of statistics can assign a number to the probability of throwing 100 heads in a row, it would never happen in the real world. Thus, while an event may be possible *statistically*, a point is reached were it must be considered to be impossible *practically*.

Mathematician, Dr. William A. Dembski, author of *"Eliminating Chance Through Small Probabilities"*, explains that chance must be discounted as a plausible explanation for an event once it requires resources greater than the available resources of a system.[21] In regard to the statistical probability of evolution having occurred by chance, the available resources are the limited time that the universe has existed (estimated by evolutionists to be 14.5 billion years, or $4 \times 10$ to the power of 17 seconds) and the limited number of atoms within the universe that are available for random chemical combinations (estimated to be 10 to the power of 80). The limited time that the universe has existed and the limited number of atoms that the universe contains are the available resources that place a constraint upon the probability of evolution having occurred. They define the upper limit of plausibility. If evolution by random processes would require more than the upper limit of available resources, then it must be regarded as implausible. We have already seen, however, that the improbabilities of even the simplest building blocks of life being formed by chance are billions or trillions of times higher than these upper limits of plausibility. Trillions of times more seconds would be needed, and trillions of times more atoms, to serve as the raw material for chance combinations, would be needed, in order for even the simplest DNA molecules to have formed by random processes. We must conclude, therefore, that it is completely impossible for even the simplest of life to have evolved by chance. Abiogenesis defies the logic of both biology and mathematics!

Recognising this impossibility, Dr. Eugene V. Koonin (not a creationist) states:

# The Origin of Life

> *"The origin of life is the most difficult problem that faces evolutionary biology and, arguably, biology in general. Indeed, the problem is so hard, and the current state of the art seems so frustrating, that some researchers prefer to dismiss the entire issue as being outside the scientific domain altogether, on the grounds that unique events are not conducive to scientific study."*[22]

In other words, Dr. Koonin is saying that *we have no idea how life could have originated; the problem is too difficult to solve, so some of us are just choosing to ignore it!* He also appears to be admitting that the answer to the origin of life must lie BEYOND the realm of science and naturalistic causes.

## MYSTERIOUS "HIGHER LAWS"

Other scientists, however, have not given up trying to account for the origin of life. In recognition of the impossibility of purely random, mechanistic forces creating life on earth, some scientists have postulated the existence of, as yet, unknown biological forces or laws that predispose the natural universe to produce biological life. According to this theory, biological life is inevitable in a universe with the right pre-existing conditions, because there are mysterious biological forces at work, which are overriding the probabilities of blind chance and steering the natural world inexorably towards the creation of life.

For example, Dr. Alexander I. Oparin, in his book, *"The Origin of Life"*, refers to the existence of *"laws of a higher order"*.[23] Similarly, Dr. Stephen J. Gould writes:

> *"Life, arising as soon as it could, was <u>chemically destined to be</u>, and not the chancy result of accumulated improbabilities."*[24]

Biochemist, Dr. Christian DeDuve, agrees with Gould:

> *"Another lesson of the Age of Chemistry is that <u>life is the product of deterministic forces</u>. Life was bound to swiftly arise under the prevailing conditions, and it will arise similarly wherever and whenever the same conditions obtain... Life and mind emerge not as the result of the freakish accidents, but as natural manifestations of matter, <u>written into the fabric of the universe</u>."*[25]

In this way, evolutionists are able to conveniently side-step the statistical and biological impossibility of abiogenesis, by claiming the existence of mysterious natural laws that have not yet been discovered. This sounds like a "Star Wars-like force" that pervades the universe; a predetermined bias of the universe to stack the deck and load the dice to ensure that life is created. But this is not science! This is merely unsubstantiated wishful thinking!

**ALIENS CREATED US**

Not all scientists, however, are comfortable with the belief in imaginary "higher laws". Instead, some have chosen to believe that life was seeded upon the earth from outer space. This theory was popularised by the noted astronomer and astrophysicist, Dr. Sir Fred Hoyle (1915 - 2001), in his book, *"Evolution From Space"*. He states:

> *"I don't know how long it is going to be before astronomers generally recognize that the combinatorial arrangement of not even one among the many thousands of biopolymers on which life depends could have been arrived at by natural processes here on the earth. ...The notion that not only the biopolymers but the operating programme of a living cell could be arrived at by chance in a primordial organic soup here on the Earth is evidently nonsense of a high order. Life must plainly be <u>a cosmic phenomenon</u>."* [26]

He goes on to say:

# The Origin of Life

> "The likelihood of the spontaneous formation of life from inanimate matter is one to a number with 40,000 noughts after it...<u>It is big enough to bury Darwin and the whole theory of evolution.</u> There was no primeval soup, neither on this planet nor on any other, and if the beginnings of life were not random, <u>they must therefore have been the product of purposeful intelligence.</u>"[27]

It is important to note that Dr. Hoyle was not a Christian, nor a believer in God. His reference to "purposeful intelligence" is a reference to his belief in aliens who seeded our planet with life. In subsequent chapters of his book, Dr. Hoyle explains his theory further. In chapter 5, "*Evolution by Cosmic Control*"[28], he argues that aliens seeded the universe with genetic material, "*cosmic genes*", allowing them to drift through space until they reached a suitable planet on which to thrive. In chapter 6, "*Why Aren't The Others Here?*"[29], he even proposes a variety of speculative reasons why these aliens have not yet visited us.

More recently, outspoken atheist, Dr. Richard Dawkins has offered a similar postulation. In an interview with American journalist and commentator, Ben Stein, Dr. Dawkins was pushed to explain how life could have originally started on earth. After conceding that there was no known natural biological process whereby life could have possibly formed, he made the following extraordinary statement:

> "It could be that, at some earlier time, somewhere in the universe, a civilisation evolved, by probably some kind of Darwinian means, to a very, very high level of technology, and designed a form of life that they seeded onto, perhaps, this planet."[30]

The idea that aliens created life on earth, however, raises two very obvious questions: Where are they? And, more importantly, who created *them*? Suggesting aliens as the originators of life on earth merely pushes back the question of ultimate origins by an additional step. It does not solve the problem of how *any* life can evolve from non-living matter *anywhere* in the universe!

**PANSPERMIA and TRANSPERMIA**

An alternate view to the alien hypothesis that is suggested by some scientists is the idea that molecules and spores of biological life exist throughout the universe, clinging to space dust, comets and meteors (*"panspermia"*). These biological spores supposedly drifted to earth from other parts of the universe, hitching a ride on space dust and planetary debris, and bursting into life again once it encountered earth's favourable conditions. This theory is known as *"transpermia"*. Some advocates of transpermia even suggest that these life spores may have originated on Mars, and that life on earth may be the second evolution of life in our galaxy after life on Mars died out.[31] Thus, humans may really be Martians who have are enjoying a second evolution of life. As Mark Wall, senior writer for the online magazine, Space.com, wrote in an article in August 2013:

> *"We may all be Martians."*[32]

But once again, the theory of transpermia begs the most obvious question: How did *that* life originate from non-living matter in the beginning?

Such an obvious objection, however, has not deterred the theory from taking root within a small but growing segment of the scientific community. Speaking at the "Origins Symposium" in April 2009, Dr. Stephen Hawking stated his opinion that life may well have been transported throughout the universe, carried within meteors.[33] In October 2018, some Harvard astronomers published a theoretical model for how dormant biological spores could be transported across the vast distances of space between galaxies.[34]

**CONCLUSION**

The theory of abiogenesis, like its predecessor, spontaneous generation, has been overwhelmingly discredited. The concept of biological life arising from dead, inanimate matter has been shown to be biologically, genetically, chemically and mathematically impossible.

Evolutionists who recognise the impossibility of abiogenesis are currently left with three alternate theories for the origin of life:

- Mysterious "higher laws" (for which we have no evidence) led to the creation of life on earth.

- Aliens (for which we have no evidence) seeded life on earth.

- Biological spores from life on other planets (for which we have no evidence) seeded life on earth.

These are wildly fanciful theories, completely devoid of evidence, conjured up from the speculative imaginations of scientists who are desperate to avoid belief in a Creator God.

---

**ENDNOTES - Chapter 9**

---

[1] Felix Archimedes Pouchet, "Heterogeny, or the treatise of spontaneous generation", Paris, 1859

[2] Wald, George. 1954. The origin of life. *Scientific American* 191:46.

[3] Oparin, A. I. "The origin of life", translation by Ann Synge. In: Bernal, J. D. (ed.), The origin of life, Weidenfeld & Nicolson, London, 1967,

[4] Oparin, A. I. "The origin of life", translation by Ann Synge. In: Bernal, J. D. (ed.), The origin of life, Weidenfeld & Nicolson, London, 1967, p. 29.

[5] https://en.wikipedia.org/wiki/Miller%E2%80%93Urey_experiment

[6] https://en.wikipedia.org/wiki/Abiogenesis

[7] Alonso Ricardo and Jack W. Szostak, *"Origin of Life on Earth"*, Scientific American, Vol. 301, Sept. 2009, p.38.

[8] Trudy McKee and James R. McKee, "Biochemistry: The Molecular basis of Life", 3rd edition, New york, McGraw Hill Publishers, 2003, p.58.

[9] G.T. Javor, Loma Linda University, grisda.org/origins/25002.htm

[10] John F. Ashton, "Evolution Impossible", Green Forest, AR, Masterbooks, 2013, p.43.

[11] John F. Ashton, "Evolution Impossible", Green Forest, AR, Masterbooks, 2013, p.43.

[12] John F. Ashton, "Evolution Impossible", Green Forest, AR, Masterbooks, 2013, p.43.

[13] John F. Ashton, "Evolution Impossible", Green Forest, AR, Masterbooks, 2013, p.43.

[14] Clifford D. Sirnak, "Trilobites, Dinosaurs and Man", New Yrork, St Martins Press, 1966, p.54

[15] Isaac Asimov, "The Genetic Code", New York, The New American Library, 1962, p.92

[16] Isaac Asimov, "The Genetic Code", New York, The New American Library, 1962, p.92

[17] Isaac Asimov, "The Genetic Code", New York, The New American Library, 1962, p.92

[18] Lawrence Lessing, "DNA: At The Core of Life Itself", New York, Macmillon Co. 1966, p.15

[19] Josh McDowell and Don Stewart, "Reasons Skeptics Should Consider Christianity", San Bernadino, CA, Here's Life Publishers, 1981, p.134

[20] https://www.physicsoftheuniverse.com/numbers.html

[21] William A Dembski, "Eliminating Chance Through Small Probabilities", Quoted in John F. Ashton, "Evolution Impossible", Green Forest, AR, Masterbooks, 2013, p.43.

[22] Koonin, Eugene V., *The logic of chance: The nature and origin of biological evolution*. Pearson Education, NJ. 2012, p.351.

[23] Alexander I. Oparin, "The Origin of Life", New York, Dover publications, 1965. p.60.

[24] Gould, Stephen J. 1990. Enigmas of the small shellies. *Natural History:* October:16-17. p.16.

[25] DeDuve, Christian. 1996. *Vital dust*. Basic Books, New York, NY. p.15.

[26] Hoyle, Sir Fred. 1981. The big bang in astronomy. *New Scientist* 92:526-527.

[27] Hoyle, Sir Fred, Wickramasinghe, Chandra. 1984. *Evolution from space*. Simon & Schuster, New York, NY., p.148.

[28] Hoyle, Sir Fred, Wickramasinghe, Chandra. 1984. *Evolution from space*. Simon & Schuster, New York, NY., Chapter 5. For a summary see: http://wasdarwinwrong.com/kortho47.htm

[29] Hoyle, Sir Fred, Wickramasinghe, Chandra. 1984. *Evolution from space*. Simon & Schuster, New York, NY., Chapter 6. For a summary, see: http://wasdarwinwrong.com/kortho47.htm

[30] Video Clip of interview between Ben Stein and Dr. Richard Dawkins, "Richard Dawkins Believes Extraterrestrials Created Man." https://www.youtube.com/watch?v=AiVoS78lNqM

[31] https://www.space.com/22577-earth-life-from-mars-theory.html

[32] https://www.space.com/22577-earth-life-from-mars-theory.html

[33] *Weaver, Rheyanne (April 7, 2009)*. "Ruminations on other worlds". *statepress.com. Archived from* the original *on July 24, 2011*. Retrieved 25 July 2013

[34] Shostak, Seth (26 October 2018). "Comets and asteroids may be spreading life across the galaxy - Are germs from outer space the source of life on Earth?". NBC News. Retrieved 31 October 2018.

Meet Your Ancestors

Chapter 10

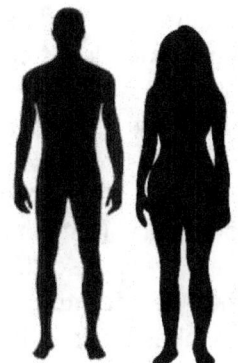

# Meet Your Ancestors

We cannot leave the topic of genetics without mentioning the fascinating controversy surrounding Mitochondrial Eve and Y-Chromosomal Adam. While these two ancient progenitors (genetic ancestors) do not necessarily disprove evolution, they fit much more neatly with the creationist's worldview than with evolution. In fact, as you will soon see, their discovery initially caused considerable angst among evolutionists, ultimately requiring them to massage their theory to fit the new facts. The resulting, modified theory of human evolution now involves some extraordinarily unlikely assumptions which seem to defy common sense.

## THE FACTS

In 1987, researchers announced that they had discovered genetic proof that all people alive today are descended from a single female. The research team, led by Drs. Allan Wilson and Rebecca L. Cann, had studied the mitochondrial DNA from women in different parts of the world and claimed that it showed that the lineage of all modern humans can be traced back to a single female progenitor.[1]

DNA is found in two locations in a cell; in the nucleus of a cell (nuclear DNA) and in the cell's mitochondria (mitochondrial DNA - mtDNA). The mitochondria is the power factory of the cell, converting the energy from sugars, fats and proteins that we eat into molecules used to power the cell. Whereas nuclear DNA is inherited from both father and mother (half of your nuclear DNA is contributed by your father and half by your mother), your mitochondrial DNA (mtDNA) comes solely from your mother.[2] (Very recent research, published in November 2018, suggests that, in rare cases, some of the father's mtDNA can infiltrate the mitochondria of offspring.[3]) By studying mutational variations in female mtDNA in different parts of the world, the early researchers were able to show that all women (and, therefore all men too) were descended from a single female in the distant past.

The research team, consisting of Drs. Allan Wilson, Mark Stoneking, Rebecca L. Cann and Wesley M. Brown, published their research paper in *"Nature"* magazine on 1st January 1987.[4] Another researcher, Dr. Roger Lewin, piggybacking off the research by Wilson et. al., published an article in *"Science"*, also in 1987, entitled *"The Unmasking of Mitochondrial Eve"*.[5] The name, *"Mitochondrial Eve"* stuck. A further 1987 article published in *"Nature"* was entitled *"Out of the Garden of Eden"*.[6]

A leading geneticist at the time, Dr. Karl Skorecki, publicly confirmed Wilson and Cann's findings:

> *"Analysis of mitochondrial DNA of all contemporary humans sampled today indicates that all of the different variations in the sequence of mitochondrial DNA (mtDNA) trace back, or converge to an original sequence in a given woman. That woman, Mitochondrial Eve, transmitted her mtDNA sequence to her offspring, and over generations slight variations in sequence occurred and accumulated, leading to the diversity of existent sequences in men and women populating the earth today."[7]*

The story soon became a world-wide media sensation. The startling news of a single female progenitor became the cover story of *"Time"* magazine, on 26 January 1987, and *"Newsweek"* printed a cover story, on 11 January 1988, with a picture of Adam and Eve, with the title, *"The Search for Adam and Eve"*.

*"National Geographic"* magazine soon followed, with the cover story, *"Mankind Meet Your Mother"*.

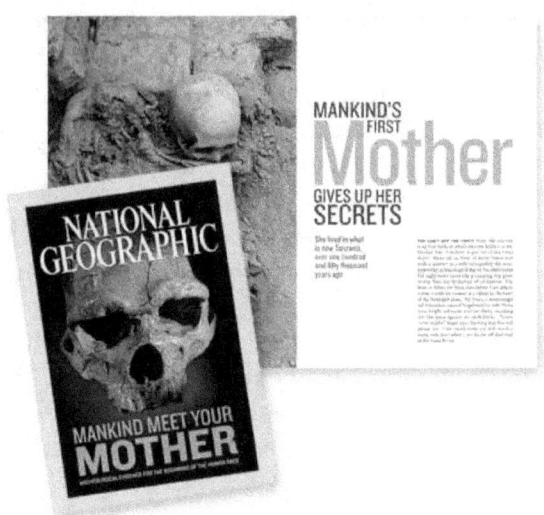

In 2002, a Discovery Channel documentary was released, entitled, "*The Real Eve: 5 Billion People From One Woman*".

Wilson and Cann's research was also able to suggest a timeline and geographical location for Mitochondrial Eve. According to evolutionary timescales, she lived somewhere in Africa between 100,000 to 200,000 years ago.[8]

**INITIAL REFUTATION**

How did evolutionists respond to this discovery? Initially, there was considerable scepticism and vehement opposition, because the discovery shattered the previous metanarrative of evolution. Until the discovery of Mitochondrial Eve, evolutionary theory had proposed that humans had gradually evolved in different parts of the world, with a large number of pre-humans gradually evolving into the antecedents of modern humans. Consequently, our DNA should have evidence of thousands of progenitors. Accordingly, Wilson and Cann's research was attacked by many leading evolutionists, and their methodology questioned. Palaeoanthropologist, Dr. Milford Wolpoff, labelled their findings as "whacko!".[9] Dr. Blair Hedges published data that seemed to contradict Wilson and Cann's findings.[10] As a result of these, and other refutations by leading evolutionists, Mitochondrial Eve seemed in serious trouble. In 1992, "*Science*" magazine published the article, "*Mitochondrial Eve: Wounded, But Not Dead Yet*", by Dr. A Gibbons.[11] In that same year, "*Newsweek*" published the article, "*Eve Takes Another Fall*", and "USA Today" announced "*anthropologists are saying it's time to write Eve's obituary*".[12] Significantly, a large part of the evolutionists counter-argument against Mitochondrial Eve was the supposed contradictory evidence in the fossil record, which seemed to indicate a diversity of human and pre-humans antecedents in different parts of the world.

At the 1993 meeting of the American Association for the Advancement of Science (AAAS), Dr. Milford Wolpoff announced, "*It's over for Eve*".[13] However, at that same meeting, Dr. Maryellen Ruvolo, from Harvard University, presented stunning new data that confirmed the original Mitochondrial Eve findings. Her studies had investigated a different part of the mitochondria, and she got exactly the same result.[14] Subsequent genetic studies have confirmed the original findings. All people alive today can be traced back to a single female progenitor. Wikipedia states:

> "Shortly after the 1987 publication, criticism of its methodology was published. Although the original publication did have analytical limitations, the findings have since proven robust."[15]

A 2010 paper, published by researchers from Rice University, Houston, Texas, stated:

> "The most robust statistical examination to date of our species' genetic links to "mitochondrial Eve", the maternal ancestor of all living humans, confirms that she lived about 200,000 years ago"[16] [more on this dating later].

Subsequent studies have been conducted on the male Y Chromosome, which is inherited patrilineally (passed on from father to son). A similar discovery has been made regarding the existence of a single male progenitor, labelled Y-Chromosomal Adam, (although he is less important genetically, because only males inherit the Y chromosome, whereas both males and females inherit their mother's mitochondrial DNA). It is now apparent that all males who are alive today can be traced back to a single male progenitor as well.[17]

**THE EVOLUTIONISTS' EXPLANATION**

How do evolutionists explain Mitochondrial Eve and Y-Chromosomal Adam? After initially ridiculing the discovery of Mitochondrial Eve, and fighting vehemently against the research data, evolutionists have now embraced the findings. In fact, Mitochondrial Eve has now become a centre-piece in evolutionary theory. Talk to most evolutionists today, and they will hail Mitochondrial Eve and Y-Chromosomal Adam as an integral part of the wonderful theory of evolution; such is their ability to place an evolutionary spin on any new evidence! For example, despite the initial vehement arguments by evolutionary palaeontologists that anthropological fossils (fossils of humans) completely contradict the concept of a single progenitor, palaeontologists are now saying *"the theory of a Mitochondrial Eve is supported by anthropological evidence."*[18] How do evolutionists manage to execute such spectacular backflips without losing credibility?

There are three major theories as to how Mitochondrial Eve fits into evolutionary history: The Bottleneck theory, the Out of Africa theory and the Coalescence theory (the latter often used in conjunction with either of the first two theories).

The Bottleneck theory proposes that a sharp reduction in population size occurred at some point in the past, precipitated by severe environmental factors such as flooding, famines, earthquakes, fires, disease or droughts (yet they deny a global flood!). This dramatically reduced the population to a very small number of survivors, who then repopulated the earth. In the case of Mitochondrial Eve, the lineages of all but one of those few female survivors died out, resulting in all people living today being descended from a single female. A female mitochondrial lineage becomes extinct when all the women from a particular lineage only produce male offspring, and can no longer pass on their mitochondrial DNA (mtDNA).

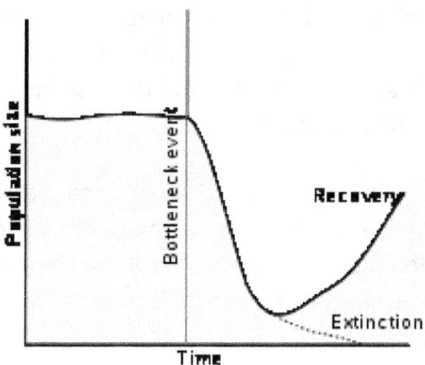

# Meet Your Ancestors

### Graph from Wikipedia.org

Given the fact that evolutionists maintain that Y-Chromosomal Adam lived at a much earlier time to Mitochondrial Eve, this means that there had to have been two such bottlenecks in our distant past. Furthermore, the Bottleneck theory requires that two global, catastrophic events must have occurred to almost completely wipe out the human population. Until the discovery of Mitochondrial Eve, evolutionists had consistently argued against the creationist story of a global flood, claiming that there is no evidence for such a global event. Now, however, adherents of the Bottleneck theory have had to execute a remarkable backflip, not, of course, in acknowledging Noah's flood, but in maintaining that a global catastrophe or a number of simultaneous catastrophes, *did* occur at some point in the past.

The Out of Africa theory is slightly different. This theory supposes that a very small group of early humans, originating in Africa, gradually increased in number, breeding only within their group and spreading across the face of the earth, eventually replacing all other humans who were living at that time. Once again, this necessitates that only one female from this African group perpetuated her lineage through to today, with the lineages of the other original African females dying out. The Out of Africa theory is widely proposed by many evolutionists today, as the best explanation of the evolution of human life that accords with the existence of Mitochondrial Eve.  A major problem with this theory, however, is that it requires these African humans to have never intermarried and bred with other humans living at the time, which is an assumption that defies common sense.

Both the Bottleneck theory and the Out of Africa theory refute the notion that Mitochondrial Eve was the first woman. They propose that she was one of many hundreds of thousands of women alive at her time, but that only her lineage has survived to this day.

A third theory, often used in conjunction with either the Bottleneck theory or the Out of Africa theory, is a genetic concept, referred to as Coalescence theory. Simply put, this means that an original population consisting of a large variety of genetic characteristics can, over time, coalesce into a single homogenous set of genetic markers. The theory proposes that the random processes of genetic recombination through sexual reproduction can eliminate all but one of the original variations, or at least eliminate any trace of its presence within the genome (DNA), causing a single original genetic code to be the only one to observably survive. Hence, Dr. Colin Mathers, from the University of Sydney, writes:

> "Mitochondrial Eve was not the first woman, nor was she the only woman alive at the time. Her contemporaries could still have descendants, whose line zig-zagged back and forth between males and females, but Eve was the only one who had an unbroken line of daughters, through thousands of generations, right down to the present time. Similarly, Y-chromosomal Adam, who did not live at the same time as Eve, was not the only man around, and not the only man with living descendants, but was the only man who every man alive today is descended from, through an unbroken line of males."[19]

To put it even more simply, a recent article on Talkorigins.org states:

> "Mitochondrial Eve is NOT our common ancestor, or even common genetic ancestor. She is the most recent common ancestor of all humans alive on earth today with respect to matrilineal descent."[20]

**THE CREATIONIST EXPLANATION**

The problem with all three theories above, is that none of the theories are possible, in a practical sense, unless the human population was reduced to a very small number of individuals at some point in the past. The best way of illustrating this is to use the analogy of inherited surnames. In traditional Western culture, surnames are inherited from fathers. A surname can become extinct if, at any point, all reproducing members of the current generation of a family are females and there are no males to perpetuate the surname into the next generation. In this case, there are still living descendants of the family, but as they are females, the surname becomes extinct. This is how evolutionists explain the existence of Mitochondrial Eve. They propose that there were many other women alive at the time, but each of their observable mitochondrial lineages died out when their reproductive offspring were all males.

The problem with this explanation, is that it is really only statistically possible if there are a very small number of original progenitors (or surnames, using the previous analogy). Additionally, the extinction of each of the lineages must happen very quickly, before too many generations have occurred. Once there are more than a handful of original progenitors, and too many generations of their offspring, it becomes statistically impossible for their lineages to die out.

For example, the more men there are with the surname "Smith", the less likely it will be none of them will have a son to perpetuate the surname. And with every additional surname added to the pool, the statistical improbability of none of them having a son increases exponentially, until it very quickly becomes a nonsense. This is why we don't all have the same surname today! The mathematics of probability and chance dictate that while *some* surnames will die out, the vast majority will flourish and multiply if there is a big enough population base. Only in a very small population base can all but one surname become extinct.

In fact, this is precisely what happened on Pitcairn Island. Dr. Carl Wieland, from Creation Ministries International, explains:

> *"Something like this happened on Pitcairn Island. Of the nine Bounty mutineers, six families settled the island in 1790. Of those six names [Christian, Young, Adams, McCoy, Quintal, Mills] only*

> the first two have survived. ... This is only probable where there is only a small number of surnames initially, i.e., a small original population; if the number of surnames is too large, it becomes very improbable for it to narrow down to only one or two."[21]

In the example of Pitcairn Island, the fact that two of the four original surnames survived shows that even in a statistically small original group of just six surnames, the chances of five out of six surnames becoming extinct is statistically very small. In a much larger population of hundreds of thousands of people, with thousands of surnames, the chances of all but one surname becoming extinct are so statistically improbable as to be rendered impossible in a practical sense.

The same is true with genetics. For the mitochondrial lineage of only one female out of hundreds of thousands to have survived until the present day, it requires that all the other lineages had to have had only male reproductive descendants at some point, which is so improbable as to effectively be a statistical impossibility.

It is in recognition of this fact, that the Out of Africa theory proposes a *very small* original group or family of African humans who gradually increased in number, breeding only within their group, and spreading across the face of the earth, eventually replacing all other humans who were living at that time. As previously noted, however, this requires the totally unbelievable assumption that these people never intermarried or bred with other humans as they spread around the globe.

So, what is the creationist explanation for the existence of Mitochondrial Eve and Y-Chromosomal Adam? While these facts can be massaged into evolutionary theory, doing so requires making some extremely implausible assumptions. On the other hand, the existence of a single male progenitor and a single female progenitor fit perfectly with the Biblical account of Adam and Eve. Some creationists initially lauded the discovery of Mitochondrial Eve, trumpeting it as certifiable "proof" of the Biblical account. This over-enthusiastic response failed to acknowledge the possibility of an alternate, evolutionary explanation, however unlikely it may be.

It must be said, however, that while Mitochondrial Eve and Y-Chromosomal Adam do not unequivocally "prove" the Bible account, they certainly provide extremely strong evidence for its veracity. Furthermore, the creationist explanation is a much simpler, neater explanation of the facts than the evolutionary explanation.

**WHEN DID MITOCHONDRIAL EVE LIVE?**

Evolutionary estimates for the dating of Mitochondrial Eve average around 200,000 years ago. This date is arrived at by estimating mutational rates of mitochondrial DNA (mtDNA) and combining this with accepted dates from the fossil record in the geological column (which we have already seen, must be viewed as completely unreliable!). However, Dr. Carl Weiland notes that two recent studies have revealed significantly faster mutation rates within mtDNA.[22] A research team, led by Dr. T.J. Parsons, published a paper in *"Nature Genetics"*, showing mtDNA mutation rates 20 times faster than current estimates.[23] The team were expecting to find a mutation rate of one in every 600 generations, but were *"stunned"* to discover a rate of one mutation in every 40 generations.[24] A review of this research, in the *"Research News"*, section of the same journal, stated that *"using the new clock, she [Mitochondrial Eve] would be a mere 6,000 years old"*,[25] a remarkable "bullseye" for the creation story!

Commenting on the response of evolutionists to this new data, Dr. Carl Wieland, states:

> *"Evolutionists have tried to evade the force of these results by countering that the high mutation rate only occurs in certain stretches of DNA called 'hot spots' and/or that the high (observed) rate causes back mutations which 'erase' the effects of this high rate. Therefore, conveniently, the rate is assumed to*

*be high over a short timespan, but effectively low over a long timespan. However, this is special pleading to get out of a difficulty, and the burden of proof is on evolutionists to sustain the vast ages for 'Eve' in the face of these documented, modern-day mutation rates."*[26]

**CONCLUSION**

- The discovery, in current human DNA, of evidence of a common, single female progenitor, Mitochondrial Eve, together with the male equivalent, Y-Chromosomal Adam, were initially disbelieved and vehemently opposed by evolutionists. This is because these new findings were inconsistent with evolutionary theory at that time.

- Once it became clear that the findings were beyond dispute, evolutionary theory was adjusted to incorporate the new data!

- The various evolutionary explanations of Mitochondrial Eve and Y-Chromosomal Adam require theoretical assumptions that are statistically highly improbable, and defy common sense.

- The discovery of Mitochondrial Eve and Y-Chromosomal Adam fit very neatly and easily with the Biblical account.

- The old evolutionary dating for Mitochondrial Eve and Y-Chromosomal Adam have been challenged by recent hard data which indicates a possible age that fits neatly into the young age of the earth and of human history, as indicated by a straight-forward reading of the Bible.

## ENDNOTES - Chapter 10

[1] https://en.wikipedia.org/wiki/Mitochondrial_Eve

[2] http://theconversation.com/study-shows-mitochondrial-dna-can-be-passed-through-fathers-what-does-this-mean-for-genetics-107641

[3] http://theconversation.com/study-shows-mitochondrial-dna-can-be-passed-through-fathers-what-does-this-mean-for-genetics-107641

[4] https://en.wikipedia.org/wiki/Mitochondrial_Eve

[5] https://en.wikipedia.org/wiki/Mitochondrial_Eve

[6] https://en.wikipedia.org/wiki/Mitochondrial_Eve

[7] Dr. Karl Skorecki, https://www.breakingisraelnews.com/97584/scientists-trace-humanity-back-single-adam-eve-strengthening-creation-story/

[8] https://en.wikipedia.org/wiki/Mitochondrial_Eve

[9] http://www.mhrc.net/mitochondrialEve.htm

[10] http://www.mhrc.net/mitochondrialEve.htm

[11] http://science.sciencemag.org/content/257/5072/873

[12] http://science.sciencemag.org/content/257/5072/873

[13] http://www.mhrc.net/mitochondrialEve.htm

[14] http://www.mhrc.net/mitochondrialEve.htm

[15] https://en.wikipedia.org/wiki/Mitochondrial_Eve

[16] https://www.sciencedaily.com/releases/2010/08/100817122405.htm

[17] http://www.sci-news.com/genetics/science-y-chromosomal-adam-01709.html

[18] https://en.wikiversity.org/wiki/Controversies_in_Science/Was_there_a_mitochondrial_Eve%3F

[19] https://www.quora.com/How-is-it-possible-that-most-Eurasian-men-can-trace-their-ancestry-back-to-only-a-dozen-ancestors-who-lived-5000-years-ago-I-am-skeptical

[20] http://www.talkorigins.org/faqs/homs/mitoeve.html

[21] https://creation.com/a-shrinking-date-for-eve

[22] https://creation.com/a-shrinking-date-for-eve

[23] Parsons, T.J. *et al* 'A high observed substitution rate in the human mitochondrial DNA control region', *Nature Genetics* **Vol. 15:** 363–368, 1997;

[24] Parsons, T.J. *et al* 'A high observed substitution rate in the human mitochondrial DNA control region', *Nature Genetics* **Vol. 15:** 363–368, 1997;

[25] *Nature Genetics* **Vol. 15:** 1997;

[26] https://creation.com/a-shrinking-date-for-eve

# A Little Common Sense Please!

## Chapter 11

# A Little Common Sense Please!

To this point, we have examined some of the major flaws in the theory of evolution. These include:

- Fossil evidence contradicting gradual evolution.

- The complete absence of transitional forms in the fossil record.

- Recent evidence contradicting previously assumed slow sedimentation rates and stratification of rocks.

## A Little Common Sense Please!

- The complete unreliability of radiometric dating of rocks and other materials

- Evidence for a much younger earth, including evidence from Carbon 14 dating.

- Evidence that dinosaurs and humans lived on the earth concurrently.

- The problem of irreducible complexity.

- The inability of evolution to account for the existence of DNA.

- The impossibility of the processes of adaptation and variation to produce completely new species.

- The impossibility of incremental positive mutations to create a new species.

- The impossibility of natural processes to create the vast amounts of new genetic information necessary for evolution to occur.

- The impossibility of abiogenesis - of non-living matter creating biological life.

These are major areas of concern that significantly undermine the credibility of the theory of evolution. In fact, for me, there is no credibility left at all. I cannot believe in a theory for which there is so much overwhelming and consistently contrary evidence.

The major areas listed above, however, are not the end of the argument. There remain dozens of contrary indicators, some of them matters of simple common sense, which further undermine evolutionary theory. This chapter presents an assortment of observations and evidence that cast further doubt upon its central tenets. Many of them deal with

## A Little Common Sense Please!

additional observational and deductive evidence for a much younger earth. If the earth is not billions of years old, then the foundation for evolution is removed entirely. Others simply challenge the gradual, uniformitarian processes upon which the theory is built.

Creation Ministries International (creation.com) has listed 101 examples of scientific evidence that contradict evolution and point to a young earth, and provides links to further reading of peer-reviewed scientific publications on many of the topics. The following is small sample of the array of evidence and common-sense observations that many creationists are proposing.

**ASTRONOMICAL EVIDENCE**

🪐 The sun consumes about 600 million tons of its mass ***every second***.[1] The ongoing nuclear reaction within the sun's core converts hydrogen to helium (a denser molecule). This results in the sun's core becoming denser, which compresses the core. In turn, this increases the fusion rate of the nuclear reaction.[2] If the sun had existed for 4.5 billion years, it should have increased in brightness and heat by about 40% as a result of this increased fusion rate.[3] This would either make it too hot today for life to be feasible, or, if we extrapolate backwards from the temperature of today's sun, it would have been too cold in the past for life to have been feasible. The current temperature and size of the sun, however, fits perfectly with a universe that is only thousands of years old.

🪐 The stars in our Galaxy, the Milky Way, are rotating at different speeds. The stars close to the centre are rotating around the galactic core much faster than the stars in the spiral arms, towards the outer edges. In fact, this is what gives the galaxy its spiral-armed appearance, as the galaxy "winds itself up", tighter and tighter, from the inside out. The difference in speeds of stars between the inside and the outside of the galaxy is so significant, that astronomers have estimated that after only a few hundred million years, the galaxy would be so tightly wound, it would have completely lost the shape of the spiral arms, and simply be a conglomerated mass of stars. Yet our galaxy is supposed to be 14.5 billion years old.[4]

Evolutionists are aware of this conundrum, referring to it as the "winding-up dilemma". Over the years many theories have been advanced to try to explain how the galaxy can be so old, yet still have perfectly defined spiral arms. None of these theories have been able to adequately explain this problem. Even the latest theory, involving "density waves" has now been discredited by recent discoveries by the Hubble Space telescope.[5]

Of course, the simplest explanation is that the universe is only a few thousand years old!

Comets are large bodies of rock and ice which have elongated elliptical orbits that extend from close to the sun to far beyond the orbits of the planets. Each time a comet approaches its perigee (closest point to the sun) large amounts of its mass are burnt off by the sun's rays. Astronomers calculate that a comet cannot last much longer than 100,000 years before it is completely consumed. Yet evolutionary theory originally taught that comets are as old as the rest of the solar system; currently believed to be about 4.5 billion years.

Evolutionists have been aware of this dilemma for many years and have advanced several theories to try to resolve it. Some theories involve explanations of how the orbits of comets may have been altered by gravitational forces from planets. Others propose the existence of an "Oort Cloud" and "Kuiper Belt"[6] - dense bodies of ice and rock supposedly beyond the orbit of Pluto, from which new comets occasionally emerge and are pulled into elongated orbits around the sun. The Oort cloud is thought to be the source of long-period comets (with very large orbits) and the Keiper Belt is thought to be the source of small-period comets.[7]

It must be stressed, however, that both these bodies remain purely theoretical at this stage. Astronomer and outspoken evolutionist, Dr. Carl Sagan, commented:

> *"Many scientific papers are written each year about the Oort Cloud, its properties, its origin, its evolution. Yet there is not yet a shred of direct observational evidence for its existence."*[8]

While the existence of comets does not provide a conclusive rebuttal of the old age of the solar system, they are a much simpler fit with the creationist model of a young universe.

Neptune and Uranus have measurable magnetic fields, yet if the solar system is billions of years old, these magnetic fields should have faded to immeasurably minute levels long ago.[9] Until the Voyager 2 spacecraft flew past these two planets in 1986 and 1989, scientists were unable to measure whether they had a magnetic field at all. Prior to the fly-bys, secular scientists made some theoretical calculations, taking into account an age of 4.5 billion years, and predicted a very small magnetic field for both planets. At around the same time (1984), acclaimed physicist and creationist, Dr. Russell Humphreys, undertook similar theoretical calculations based on the premise that the solar system was only about 6,000 years old.[10] He arrived at a calculated magnetic field for both planets that was **100,000 times greater** than the figure proposed by evolutionists.[11] This was treated with ridicule by evolutionists, until the Voyager 2 spacecraft flew past both planets and measured their magnetic fields. The result? Their magnetic fields were almost exactly as Dr. Humphreys had predicted![12]

## A Little Common Sense Please!

Evolutionists are left with a conundrum: How can two supposedly ancient, cold planets have such a strong magnetic field? Some scientists have attempted to explain this puzzle by speculating about possible variations in the decay rate of dipole-field strengths in the past.[13] However, the explanation that best fits the facts is that the universe is only thousands of years old.

A similar incident occurred in 1975. The Mariner 10 space probe made three fly-bys of Mercury and measured a surprisingly strong magnetic field.[14] Secular scientists were perplexed, because such a strong magnetic field is not consistent with an age of billions of years. If Mercury is as old as evolutionists maintain, it would long ago have cooled down enough to solidify its core, thus preventing the "dynamo" mechanism that secular scientists propose as the driving force that produces planetary magnetic fields. Of course, the strength of the magnetic field fits perfectly with the creationist view of a solar system that is only thousands of years old.

Following the discovery of Mercury's magnetic field, evolutionists calculated that Mercury's magnetic field is diminishing at a much slower rate than previously thought, thus explaining its current high levels. Creationists, however, argued that it is simply further evidence of a very young solar system.

In 1984, Dr. Russell Humphreys calculated a much faster degradation rate of Mercury's magnetic field, based upon an estimated age of 6,000 years.[15] When NASA announced a planned fly-by of Mercury by the space probe "Messenger", to take place in 2008, Dr. Humphreys predicted that Mercury's magnetic field would have diminished by 4% from the 1975 readings.[16] This prediction was treated with ridicule by evolutionists. But the 2008 fly-by of Mercury, by the Mariner space-probe, confirmed Dr. Humphrey's prediction and contradicted the much slower degradation rates proposed by evolutionists.[17]

## A Little Common Sense Please!

Once again, the simplest explanation of these facts is that the solar system is only thousands of years old.

🪐 Evidence of recent volcanic activity on Earth's moon does not accord with an ancient age. A report in Science magazine, in 2014, stated:

> "The moon, thought to be cold and dead, is still alive and kicking—barely. Scientists have found evidence for dozens of burps of volcanic activity, all within the past 100 million years—a mere blip on the geologic timescale. And they think that future eruptions are likely—although probably not within a human lifetime."[18]

If the solar system was **billions** of years old, the moon would long ago have cooled off and no signs of recent volcanic activity would be observed.[19]

🪐 Supernova are massive stars that have exploded at the end of their life-cycle. The remnants of these explosions are calculated to keep expanding for hundreds of thousands of years, in accordance with the laws of physics. If the universe is billions of years old, we should be able to see old, widely expanded remnants of supernova explosions. However, there are **no** old, widely expanded remnants of supernova explosions (Stage 3 supernovas) in any of the observable galaxies, and there are only a few recent (Stage 1) supernova remnants.[20] This is consistent with a universe that is only thousands of years old, and *inconsistent* with a 14.5 billion-year-old universe.[21]

## A Little Common Sense Please!

The exploration of Saturn and its moons by recent space probes has led to a discovery that has perplexed secular scientists. Titan, Saturn's largest moon, has an atmosphere that is rich in methane, yet the sun's UV radiation, even at that distance, should have broken down the methane long before now. Smaller traces of methane have also been discovered in the atmosphere of Mars. In fact, scientists estimate that without biological life producing a steady supply of methane, the methane in the atmosphere of a moon the size of Titan should only last for a maximum of 10,000 years.[22] An article published in "Scientific American" by Dr. Sushil K. Atreya in 2009 stated:

> "The presence of methane in the atmospheres of Mars and Titan is one of the most tantalizing puzzles in our solar system."[23]

If the solar system is billions of years old, there should be no methane left in the atmospheres of these planets.

Furthermore, astrogeologists point out that if Titan is billions of years old, the methane in its atmosphere should have produced an ocean of ethane on the surface of the moon. There is no such sea, however, once again indicating a very young age for Titan. Puzzled evolutionist, Dr. Jonathon Lunine, comments:

> "If the chemistry on Titan has gone on in steady-state over the age of the solar system, then we would predict that a layer of ethane 300 to 600 meters thick should be deposited on the surface." No such sea is seen, which is consistent with Titan being <u>a tiny fraction of the claimed age of the solar system</u>."[24] (Underline added)

A Little Common Sense Please!

In 1979, the Voyager space probe discovered active volcanoes on Io, a moon of Jupiter.[25] The Galileo probe, in 1995, recorded at least 80 active volcanoes on Io and a sea of magma beneath its surface.[26] This was a truly astonishing discovery. If Io is billions of years old, all activity such as this should have ceased billions of years ago! If evolutionists insist on maintaining an age for Io of billions of years, it must be pointed out that even at 10% of its current rate of volcanic activity it would have erupted its entire mass 40 times over during its lifetime. In every way possible, Io presents as a very recent creation.

**HUMAN HISTORY**

In 1810, the earth's population was approximately one billion people. Two centuries later it had grown to *six* billion people. If the Bible's chronology is accurate, and the earth's population grew from just a few people after Noah's flood, about 4,400 years ago, then today's population, with its current growth rate, fits perfectly. If, however, the earth is 4.5 billion years old, and mankind has existed for about 200,000 years, the question has to be asked, "Where are all the people?". [27] If we assume an annual growth rate of only 0.5% per annum (about one third of the world's current growth rate), after 200,000 years of human history there would be trillions of people on our earth. There would not even be standing room!

Evolutionists claim that the stone age lasted about 100,000 years, with a steady population of Neanderthal and Cro-Magnon humans of between one and ten million.[28] This would mean that at least four billion

241

bodies would have been buried, along with their stone age tools, during that period. Where are all the bodies? Only a few thousand skeletons have been found from that epoch, yet we should be able to find millions. The paucity of skeletal remains from that period seems to indicate that the stone age was much more recent and lasted only a few centuries.

Evolutionists claim that mankind existed for nearly 200,000 years before agriculture was developed, only 10,000 years ago.[29] Surely this defies common sense! Surely humans would have worked out how to plant seeds and grow crops much sooner that this! This is another evidence that casts doubt upon the supposed hundreds of thousands of years of human "pre-history".

**BIOLOGICAL EVIDENCE**

Multiple discoveries of dinosaur blood cells, blood vessels, soft tissue and DNA refute the ancient ages proposed for them. These discoveries have now been found in locations in various parts of the world, and are beyond dispute.[30] This has already been discussed in detail in Chapter 6.

In 2001, bacteria was discovered in salt inclusions embedded in a supposedly 250-million-year-old strata. Despite this apparent age, Dr. M. Oard published data verifying that the bacteria was successfully revitalised, yet bacteria cannot possibly survive intact for more than a few thousand years at the most.[31]

# A Little Common Sense Please!

⚛ The existence of Carbon-14 in coal, defies the evolutionary age of millions of years. Many coal samples have now been studied, and Carbon-14, which can only last a few thousand years before it completely decays, has been consistently found.[32]

⚛ Carbon-14 in oil suggests an age of thousands of years rather than millions.

⚛ Carbon-14 in diamonds suggest an age of thousands of years rather than millions.

⚛ The observed rapid accumulation of mutations in the human genome reveals that our DNA is deteriorating at an alarming rate. The claim by evolutionists that our genome developed gradually over millions of years, growing incrementally more complex, ignores the observable, measurable tendency of DNA to deteriorate faster than it evolves. Evolution teaches that the eventual appearance of humans upon the earth was the result of billions of years of the gradual and increasingly complex evolution of biological DNA. Yet Cornell University geneticist, Dr. John Sanford, has published data indicating that our DNA accumulates 90,000 errors every 6,000 years.[33] By 6 million years, one codon in every 33 would be completely inoperable, and it is inconceivable that the code could still function at that point. In other words, humans and supposed pre-humans could not possibly have been evolving for 6 million years, because our DNA would be completely inoperable by now!

# A Little Common Sense Please!

Chemical and biological evolution requires that life has evolved into increasingly more complex and organised structures. Yet this is a complete contradiction of the second law of thermodynamics. Explained colloquially, this law means that in any ordered system, open or closed, there exists a tendency for that system to decay to a state of disorder, which can only be suspended or reversed by the application of an external source of energy and organisation. To put it even more simply, the natural tendency of all life and matter is for it to decay and deteriorate. Everything naturally falls apart.

Those evolutionists who claim that, given enough time, random processes could create ever more complex chemicals and, ultimately, organisms, are ignoring this fundamental law of science. Dr. George Wald describes this huge contradiction in regard to the issue of the origin of life:

> *"In the vast majority of the processes in which we are interested, the point of equilibrium lies far over towards the side of dissolution. That is to say, spontaneous dissolution is much more probable and hence proceeds much more rapidly than spontaneous synthesis."*[34]

As Dr. Arthur Eddington once commented:

> *"If the theory [evolution] is found to be against the second law of thermodynamics I can give you no hope; there is nothing for it but to collapse in deepest humiliation."*[35]

## GEOLOGICAL EVIDENCE

In the 1970s, creationist physicist, Dr. Thomas Barnes noted that measurements of the earth's magnetic field since 1835 reveal that the field is decaying at the rate of 5% every century.[36] Archaeological measurements also confirm this, revealing that the earth's magnetic field was 40% stronger in 1000 AD than it is today.[37] Based upon these measurements and observations, Dr. Barnes calculated that the earth cannot be more than 10,000 years old, otherwise the original strength of the magnetic fields would have been large enough to melt the earth![38]

Evolutionists argued against this, by proposing that the earth's magnetic field has reversed every few million years, thus resetting the magnetic field. Thus, they argue that Dr. Barnes' linear regression equation is invalid.

More recent developments, however, while confirming the possibility of magnetic field reversals, have shown that these could **not** halt the decay pattern. Rather, the total magnetic field would decay even faster![39]

For the evolutionists, it's a case of back to the drawing board. The rapidly diminishing magnetic field of the earth is a major problem for an ancient age of the earth, and therefore, for the theory of evolution itself. In 2014, evolutionist and geophysicist, Dr. David Stevenson, commented:

> *"Right at this moment, there is a problem with our understanding of Earth's core and it's something that's emerged only over the last year or two. The problem is a serious one. We do not now understand how the Earth's magnetic field has lasted for billions of years. We know that the Earth has had a magnetic field for most of its history. We don't know how the Earth did that. We have less of an understanding now than we previously thought*

*we had a decade ago of how the Earth's core has operated throughout history."*[40]

⛰ Our oceans contain approximately 3.6% dissolved salt and other minerals. Scientists estimate that 457 million tons of salt are added to our oceans each year from rivers and land run-off, while only 122 million tons are removed via various processes.[41] Thus only approximately 27% of added salt is removed each year. If our earth was millions or even billions of years old, the oceans would long ago have reached salt saturation. The current level of 3.6%, however, sits perfectly with an earth that is only 6,000 years old and which experienced a global flood about 4,400 years ago.[42]

⛰ The natural radioactive decay of all igneous rocks on earth produces helium as a byproduct, which is added to the earth's atmosphere. The rate of loss of helium into space is extremely negligible. If the earth is 4.5 billion years old, as evolutionists propose, there would be at least 2,000 times the amount of helium in the earth's atmosphere! (There is only 0.05% the expected amount of helium).[43]

Furthermore, a study published in the Journal of Geophysical Research in 1982, revealed that there is far too much helium still trapped in rocks if they are supposed to be billions of years old. The retention rate of helium within the rocks, noted in the article, suggests an age of only thousands of years.[44]

⛰ Wind and water continually erode soil from our continents and deposit it as sediments and mud on the ocean floors, at the rate of about 25 billion tons per year.[45] Approximately 1 billion tons of this build-up of sediment is removed by tectonic plates sliding under one another (subduction), taking some sediment deeper into the earth's crust.[46] The remaining 24 billion tons accumulates.

The average depth of sediment on our ocean beds today, is 400 meters. Yet if the earth is 4.5 billion years old there should be 375 times that amount of sediment! This equates to 150 kilometres, which, of course is completely nonsensical. In other words, our oceans would have completely silted up and the continents would be flooded if the earth was billions of years old. On the other hand, 400 metres of sediment equates to about 12 million years worth of sedimentation at current rates, a fraction of the age of the earth proposed by evolutionists. However, when we factor in a global flood that would have produced massive erosion and sedimentation, 400 metres sits very comfortably with an age of 6,000 years.

The current erosion rate of continents is another major problem for evolution and the theory of the earth's ancient age. Continents erode so rapidly that they should have worn away completely many times over if the earth is billions of years old.[47] Evolutionists estimate that the continents are 2.5 billion years old. The current vertical erosion rate of continents is measured at 6.0 mm (0.24 inches) downwards every 100 years.[48] At this rate 150 kilometres (93 miles) in height of the continental land mass would be eroded after 2.5 billion years! In other words, if the earth was as old as evolutionists maintain, there would be no continents left on earth. For example, Dr. Tas Walker calculates that:

> *"North America should have been levelled in just 10 million years if erosion has happened at the average rate. Note that this is an upper age limit, not an actual age."*[49]

# A Little Common Sense Please!

Dr. Sheldon Judson proposes a slightly slower erosion rate over time, of 2.4 mm per 100 years. Yet even at this rate, he states:

> *"At this rate the ocean basins would be filled in 340 million years. The geological record indicates that this has never happened in the past... Furthermore, at present rates of erosion, the continents, which now average 875m in elevation, would be reduced to sea level in about 34 million years."*[50]

▲ Coastal erosion rates reveal similar problems for evolution. A recent study in Great Britain measured the coastal erosion rate of the white cliffs of Dover as 30 cm (12 inches) per year.[51] At this rate Great Britain, which is approximately 437 km wide at its widest point, would be completely disappear in 1 million years. Erosion rates in the USA and Australia are observed at a higher rate of around 0.4 metres per year.[52] At that rate, the whole of Australia, for example, would erode into the sea in only five million years, **or 35 times over**, in the supposed time since the continents separated. Simply put, coastal erosion rates turn the proposed evolutionary age of billions of years into a complete nonsense.

▲ Polystrate fossils (for example, fossilised trees extending through multiple rock strata) could not have remained exposed for millions of years while the rock strata gradually formed around them. Wind, rain and biological degradation would have completely eroded them long before they could be completely buried. Polystrate fossils, as evidence of rapid sedimentation and a young earth, were discussed in detail in Chapter 5, "Dirt, Rocks and Water".

## A Little Common Sense Please!

While Darwin predicted that the fossil record would eventually reveal an abundance of transitional fossils, even 150 years later all we have are a handful of disputable examples. The lack of uncontested transitional forms in the fossil record, together with the sudden appearance of fully developed new forms, is a direct contradiction of what we should expect to find if evolution was true. Evolutionist, Dr. E.C. Olson concedes this disturbing fact:

> "New groups of plants and animals suddenly appear apparently without any close ancestors... This aspect of the record is real, not merely the result of faulty or biased collecting."[53]

**CONCLUSION**

The nature of science is such that no theory is ever proved in an ultimate sense. Even uncontested theories such as the law of gravity must leave themselves open to the possibility that they may, one day, be found to operate differently, under unique conditions that had not been previously imagined, and must, therefore be amended. Proof, in an ultimate sense, is a very rare luxury. Disproof, on the other hand, is usually relatively easy to come to. A theory which is continually contradicted by observable evidence cannot be said to be proven. It must either be amended until it fits the evidence, or, if the contradictions are too significant and too many, it must be considered to be disproved. Disproof tends to become evident reasonably quickly, provided that those investigating the matter have an open mind and are honestly seeking the truth.

None of the individual observations and evidences briefly highlighted in this chapter disprove evolution. Many of the issues raised above have been addressed by evolutionists who have proposed theories explaining how these anomalies can be reconciled to evolutionary timescales and processes. Their theories often require acceptance of unprovable and unlikely assumptions, as well as a degree of imagination and, for me, at times, a suspension of common sense. On the other hand, the creationist viewpoint is a much simpler, logically consistent explanation for the issues that have been highlighted.

The observations in this chapter may not constitute absolute disproof when considered **individually**. But considered **together**, they provide a very convincing body of evidence that the evolutionary timescale and processes are false. Furthermore, when the major problems that have been discussed in previous chapters are added to the mix, such as the complete failure of the fossil record to confirm gradual evolution, the impossibility of abiogenesis, the irreducible complexity of microbiological entities and the insurmountable problems posed by genetics, the body of evidence against evolution becomes overwhelming. One must ask, therefore, how much contrary evidence there needs to be before a theory can be considered to be disproved. Wherever that tipping point is, surely we have gone way beyond it now in regard to the theory of evolution.

---

**ENDNOTES - Chapter 11**

---

[1] https://socratic.org/questions/how-much-mass-does-the-sun-consume

[2] https://rationalwiki.org/wiki/101_evidences_for_a_young_age_of_the_Earth_and_the_universe

[3] https://creation.com/age-of-the-earth

[4] https://creation.com/evidence-for-a-young-world

[5] D. Zaritsky *et al.*, *Nature*, 22 July 1993. *Sky & Telescope*, December 1993, p. 10.

[6] https://solarsystem.nasa.gov/solar-system/oort-cloud/overview/

# A Little Common Sense Please!

7 IBID

8 Sagan, C. and Druyan, A., 1985. *Comets*, Random House, New York, p 201.

9 https://creation.com/young-universe-evidence

10 D. Russell Humphreys, The creation of planetary magnetic fields, *Creation Research Society Quarterly* **21**(3):140–149, 1984

11 D. Russell Humphreys, The creation of planetary magnetic fields, *Creation Research Society Quarterly* **21**(3):140–149, 1984

12 https://creation.com/the-earths-magnetic-field-evidence-that-the-earth-is-young#planets

13 https://creation.com/the-earths-magnetic-field-evidence-that-the-earth-is-young#planets

14 https://nineplanets.org/spacecraft.html

15 https://creation.com/mercurys-magnetic-field-is-young#20120328

16 https://creation.com/mercurys-magnetic-field-is-young#20120328

17 https://creation.com/mercurys-magnetic-field-is-young#20120328

18 https://www.sciencemag.org/news/2014/10/recent-volcanic-eruptions-moon

19 DeYoung, D.B., Transient lunar phenomena: a permanent problem for evolutionary models of Moon formation, *Journal of Creation* **17**(1):5–6, 2003; creation.com/tlp; Walker, T., and Catchpoole, D., Lunar volcanoes rock long-age timeframe, *Creation* **31**(3):18, 2009.

20 Davies, K., Distribution of supernova remnants in the galaxy, Proc. 3rd ICC, pp. 175–184, 1994.

21 Sarfati, J., Exploding stars point to a young universe, Creation 19(3):46–48, 1997

22 https://creation.com/focus-273-creation-magazine

# A Little Common Sense Please!

[23] https://www.scientificamerican.com/article/methane-on-mars-titan/

[24] https://www.astrobio.net/?option=com_retrospection&task=detail&id=1478

[25] https://creation.com/age-of-the-earth

[26] https://phys.org/news/2011-05-galileo-spacecraft-reveals-magma-ocean.html

[27] https://creation.com/age-of-the-earth

[28] Deevey, E.S., The human population, *Scientific American* **203**:194–204, September 1960

[29] https://creation.com/age-of-the-earth

[30] Wieland, C., Dinosaur soft tissue and protein—even more confirmation!, 2009.

[31] https://creation.com/age-of-the-earth

[32] What about carbon dating? *Creation Answers Book* chapter 4.

[33] John F. Ashton, "Evolution Impossible", Green Forest, AR, Masterbooks, 2013, p.132.

[34] George Wald, "The Origin of Life", Scientific American, Vol.

[35] Arthur Eddington, "The Nature of the Physical World", New York, MaMillan, 1930, p.74.

[36] K.L. McDonald and R.H. Gunst, 'An analysis of the earth's magnetic field from 1835 to 1965,' *ESSA Technical Report, IER 46-IES 1*, U.S. Govt. Printing Office, Washington, 1967.

[37] R.T. Merrill and M.W. McElhinney, *The Earth's Magnetic Field*, Academic Press, London, pp. 101–106, 1983

[38] https://creati191, 1954, p.49on.com/the-earths-magnetic-field-evidence-that-the-earth-is-young

[39] D.R. Humphreys, Reversals of the earth's magnetic field during the Genesis Flood, *Proceedings of the First International Conference on Creationism*, Creation Science Fellowship, Pittsburgh, **2**:113–126, 1986. / ALSO: R.S. Coe, M. Prévot and P.

Camps, New evidence for extraordinarily rapid change of the geomagnetic field during a reversal, *Nature* **374**(6564):687–692, 1995; see also A. Snelling, The principle of 'least astonishment', *Journal of Creation* **9**(2):138–139, 1995.

[40] Cited in: Folger, T., Journeys to the Center of the Earth: Our planet's core powers a magnetic field that shields us from a hostile cosmos. But how does it really work? *Discover*, July/August 2014.

[41] https://creationtoday.org/evidence-for-a-young-earth/

[42] Austin, S.A. and D.R. Humphreys, The sea's missing salt: a dilemma for evolutionists, *Proceedings of the 2nd International Conference on Creationism*, Vol. II, Creation Science Fellowship, 1991, icr.org/article/sea-missing-salt.

[43] Vardiman, L., *The Age of the Earth's Atmosphere: A Study of the Helium Flux through the Atmosphere*, Institute for Creation Research, 1990, P.O.Box 2667, El Cajon, CA 92021.

[44] Gentry, R. V. *et al.*, Differential helium retention in zircons: implications for nuclear waste management, *Geophysical Research Letters* **9**:1129–1130, October 1982. See also ref. 20, pp. 169–170.

[45] Gordeyev, V.V. *et al.*, The average chemical composition of suspensions in the world's rivers and the supply of sediments to the ocean by streams, *Doklady Akademii Nauk SSSR* **238**:150, 1980.

[46] Hay, W.W., et al., Mass/age distribution and composition of sediments on the ocean floor and the global rate of subduction, Journal of Geophysical Research93(B12):14,933–14,940, 10 December 1988.

[47] . Walker, T., Eroding ages, Creation 22(2):18–21, 2000; creation.com/erosion

[48] Roth, A., Origins: Linking Science and Scripture, Review and Herald Publishing, US, p. 271, 1998, cites Dott and Batten, Evolution of the Earth, McGraw-Hill, US, p. 155, 1988, and a number of others.

[49] Dr. Tas Walker, "Eroding Rates", https://creation.com/eroding-ages

[50] Sheldon Judson, "Erosion of the Land or What's Happening to Our Continents", American Scientist, Vol. 56, 4, p.371

# A Little Common Sense Please!

[51] https://www.forbes.com/sites/davidbressan/2016/11/12/coastal-erosion-is-accelerating-at-worrying-rate-and-human-activity-is-to-blame/#363b2ab056cd

[52] https://agupubs.onlinelibrary.wiley.com/doi/abs/10.1029/EO064i035p00521

[53] E. C. Olson, "The Evolution of Life", New York, Mentor Books, 1965, p.94.

Chapter 12

# Science and Faith

The complete failure of evolutionary theory to provide a plausible explanation for the origin of biological life and of the universe itself, leads very naturally to a discussion of the creationist worldview. By this, I mean the concept that the universe and everything in it is the product of special creation by an all-powerful, transcendent, intelligent designer. In the next few chapters we will examine the overwhelming scientific evidence for the existence of a Creator God.

But before we do, there is a popular misconception that must be addressed; the idea that faith and science do not mix. According to outspoken atheist and evolutionist, Dr. Richard Dawkins:

# Science and Faith

> *"Not only is science corrosive to religion; religion is corrosive to science. It teaches people to be satisfied with trivial, supernatural non-explanations and blinds them to the wonderful real explanations that we have within our grasp. It teaches them to accept authority, revelation and faith instead of always insisting on evidence."*[1]

According to Richard Dawkins and many like him, science is the realm of facts, evidence and truth, whereas faith is the realm of unsubstantiated, unverifiable superstition and myths. Thus, science and faith are the antithesis of each other: like matter and anti-matter, they cannot exist together. According to this view, a person cannot subscribe to the realms of faith and science simultaneously: one must choose one or the other.

Regarding the relationship between faith and science, there are two ill-informed views:

## 1. The False View That Science and Faith are at War

This is a much more common view today than it was in the past, partly because of the advent of the new militant atheism. However, as we shall see shortly, many of the great scientists of the past were people of strong Christian faith, who saw no conflict between their belief in God and their scientific studies. Much of the currently perceived hostility between science and faith has arisen because of the aggressive and very public campaigns of militant atheists such as Richard Dawkins, Christopher Hitchens and Bill Nye, who have proactively sought to belittle religious faith and portray it as unscientific. They deliberately portray faith as the enemy of science, and depict those who have faith as being fools.

However, this view of the inherent antipathy between science and faith is directly contradicted by the existence of many scientists of faith. There are many eminent and respected scientists today who are anything but fools, and they would strongly disagree that their faith is in conflict with their scientific endeavours. In fact, many would say that their faith actually enhances their scientific study of the natural world. For example, astrophysicist Dr. Jennifer Wiseman, the discoverer of the Wiseman-Skiff comet in 1987, states:

*"By studying nature, I enrich my understanding of God."*[2]

## 2. The False View That Science and Faith are Completely Unrelated

This false view sometimes goes hand in hand with the first. According to this view, science and faith deal with completely separate realms, and as long as they are kept separate, no harm is done. Thus, science deals with the natural world, whereas faith deals with the supernatural world. As long as faith remembers its place, and does not attempt to cross over into the domain of science, it can be tolerated. But if it starts to impinge upon the natural world, and influence our interpretation of the scientific evidence, it is out of line and must return to its corner. Faith must not seek to interfere with scientific endeavours, and must remain subservient to science when it comes to interpreting the observable natural world.

This attitude can manifest itself in a benign form and a more judgmental, condescending form. The benign form states; *both the natural and supernatural realms are valid but must be kept separate*. In other words, the two realms are distinct and must have no interconnection. This dualist view of the universe, however, is not supported by the Bible. If there is a God who has created the universe, he is not completely separate from it. He did not create the physical universe and then abandon it to return to his supernatural realm. The Bible portrays God as transcendent (above and beyond the created universe), but also intimately involved with his universe, and with mankind in particular. The physical and the spiritual are overlapping realms. People are physical and spiritual beings at the same time. The natural world continues to be sustained and energised by the life-giving Spirit who created it. To separate the physical from the spiritual, the natural from the supernatural, is to make a distinction that the Bible simply does not make.

The more judgmental, condescending form of the view of the incompatibility of science and faith states; *belief in a supernatural realm is superstitious nonsense, but as long as it keeps its distance, we don't mind.* An example of this more condescending form is evident in an article entitled *"Science and Faith are Not Compatible",* by Sean Carroll in *"Discover Magazine",* published on 23rd June 2009:

> *"The incompatibility between science and religion doesn't mean that a person can't be religious and be a good scientist. That would be a silly claim to make. There is no problem at all with individual scientists holding all sorts of <u>incorrect beliefs</u>."*[3]

What an extraordinarily condescending comment to make! According to this view, you can still be a good scientist and have a religious faith - as long as you leave your faith at the door when you get to work, along with your belief in Santa Claus and the Easter Bunny. Underlying this view is the presupposition that the supernatural realm is not "real", and that science has somehow proved it thus. This brings us to a discussion of the fundamental nature and limits of science, because in issuing these kinds of judgments upon the realm of faith, science has completely overstepped its own boundaries.

**THE FUNDAMENTAL NATURE AND LIMITS OF SCIENCE**

The British Science Council, a body that oversees the standards for professional registration of practising scientists across all disciplines, defines science in the following way:

> *"Science is the pursuit and application of knowledge and understanding of the natural and social world following a systematic methodology based on evidence."*[4]

It then lists the major methodologies that define the practice of science, which include:

- Objective observation and measurement of data.

- Induction (formulating theories to explain that data).

- Repetition (the ability to replicate and verify what is theorised).

- Verification (testing and critiquing a theory against results to establish its validity).

In particular, please note that science is the study of evidence and data that is **observable**, **measurable** and **repeatable**. The study of gravity is a good example. Isaac Newton and others **observed** the effects of gravity. They then **measured** its effect and, through **induction**, calculated a formula to precisely describe its action. Then through **repeated** experiments, they **verified** their formula as an accurate description of gravity's operation.

Let me reiterate: Science is the study of evidence and data that is **observable**, **measurable** and **repeatable**. Immediately, this places limits on what science can and cannot comment on. Pure, experimental science cannot claim to be able to verify, or otherwise, anything that is not observable, measurable and repeatable. It simply cannot comment on things that are intangible. For example, science cannot make any comments on the existence and nature of love. It may measure and analyse physiological responses of people who *claim* to experience love - elevated heart rate, the production of endorphins, increased metabolic rates, changes in brain activity - but it cannot measure love itself, nor can it offer any judgment as to whether love actually exists. Because love is intangible. Love cannot be measured, weighed, poked, prodded or photographed. Thus, all science can say is that it can observe and measure physiological phenomena in the natural world that are consistent with the existence of love, but it cannot offer any form of ultimate proof of love's existence.

Exactly the same is true of science's relationship to faith and to the possible existence of God. The Mirriam-Webster Dictionary defines God as:

> "The incorporeal [non-physical] divine Principle ruling over all as the eternal Spirit"[5]

While I take exception to the description of God as being a "Principle" (because he is a personal, all-powerful **being**, not merely a **principle**), this dictionary definition does at least accord with the Bible's description of God as being *"incorporeal"* (without a physical form) and an *"eternal spirit"*. If we accept these descriptions of God, then science is immediately ruled out of offering *any* kind of ultimate judgment regarding either God's existence or his nature. If an *"incorporeal, eternal Spirit"* exists, he lies completely outside the parameters of scientific speculation. He cannot be measured, weighed, poked, prodded or photographed. Science can neither prove nor disprove God's existence.

On the other hand, if a Creator God exists, it is entirely reasonable to expect to find traces of his creative handiwork in the fabric of the physical universe. If the universe was created by an all-powerful, intelligent designer, albeit an intangible, supernatural one, we should expect to be able to observe evidence of his creative design and intervention within the natural realm. Just as in the case of something intangible such as love, while God himself cannot be observed and measured, it should be possible to observe his *effects*.

Science, therefore, cannot offer any authoritative declarations on the existence or otherwise of intangibles such as love and God. All science can do is comment on observed effects that these two intangibles may produce in the natural world. To go beyond this, in declaring either absolute proof or disproof, is to exceed the parameters of science itself.

Thus, when people such as Sean Carroll, in his article *"Science and Faith are Not Compatible"* (quoted above) comments that a scientist who has a faith in God has an *"incorrect belief"*, he has stepped completely outside the defined parameters of science and made a judgment that science is not qualified to make. Science is simply not equipped to make absolute, definitive judgments about the existence of something outside of the natural world.

When a scientist says that it is unscientific to believe in God, it is he or she who is being unscientific. A scientist may profess to disbelieve in God, but to appeal to science as the basis for that unbelief makes as much sense as someone saying that he does not believe in microbes because he cannot see them with his binoculars. Binoculars are entirely the wrong instrument for finding microbes, and science is entirely the wrong instrument for finding God. Microbes will be completely invisible to binoculars, as will the transcendent God be to science (in the sense of ultimate, irrefutable proof). Both exist beyond the scope of their respective instruments to detect.

**The Inability of Science to Prove Non-Existence**

Because science is limited to studying only those things that can be observed and measured, it cannot prove that something does ***not*** exist. In this sense, it cannot prove an absolute, universal negative. All science can say is that, to date, a certain thing has not been observed. Take, for example, the possible existence of aliens. Some scientists believe that intelligent aliens may exist somewhere in the universe. In fact, there exists a scientific research program which is attempting to locate intelligent life elsewhere in our galaxy; SETI (the Search For Extra-Terrestrial Intelligence). To date, they have not located any signs of alien life. If, at some point in the future, some kind of contact is made - transmissions are received, or visual confirmation is gained - then it can be said that science has proved that aliens DO exist. Science can prove a positive.

But suppose astronomers search the cosmos in vain for a thousand years without any trace of extra-terrestrial life. Can science then authoritatively declare, *"We have proved that aliens do not exist"*? No. Because the very next day an alien spaceship may land on Bondi Beach and the crew emerge for a summer holiday. All science can say is, *"We have not yet observed aliens"*. Science cannot prove the universal non-existence of something; it cannot prove a universal negative, in this sense. I say, "in this sense", because some universal negatives can be proven if they are defined by very specific parameters. For example, the statement, *"No bachelors are married"*, is a universal negative that is entirely provable, because the conclusion is implicit in the predicate. Similarly, the statement, *"There are no full-blood Inuit Indians enrolled at Epping Boys High School this year"* is also provable (or disprovable), via examination of the enrolment register and DNA testing. In this case, the means of either proving or disproving the negative are accessible and achievable.

But science can never claim, *"aliens do not exist"*, because the proof of that universal negative would necessitate examining every corner of the universe - every planet, asteroid, rock and star - with every possible detection mechanism. I am not saying that I believe in aliens! I am simply pointing out that, in this example, the means of proving a universal negative are well beyond the reach of science.

In the same way, science can never claim, *"God does not exist"*, because the means of proving that claim is also beyond the reach of science. All science can say is *"We have not yet observed God"*. In the case of aliens, of course, a thousand years of fruitless searching should lead scientists to believe that aliens *probably* don't exist, because physical beings inhabiting the physical universe are well within the scope of science to reasonably detect. If aliens have not been detected after a thousand years of searching among the galaxies that are closest to us, there is a very strong probability that they don't exist anywhere. In the case of God, however, scientists should not be *at all* surprised at their inability to detect him, as he lies *completely* beyond the scope of their ability to perceive him.

It is truly surprising, therefore, when scientists who continue to believe in aliens, despite a lack of observable evidence, are regarded as reasonable, rational people, whereas scientists who continue to believe in God, with the support of some fairly convincing observable evidence (as we shall see in the next chapter), are often denigrated and even persecuted. The 2008 documentary film, "*Expelled: No Intelligence Allowed*",[6] reported the cases of several scientists who were sacked from their positions simply because of their belief in intelligent design.

Given these two limitations of science, its inability to prove a negative and its inability to study anything beyond the natural world, it is truly remarkable that many scientists feel competent to make authoritative declarations about the non-existence of God. Dr. Richard Dawkins' book, "*The God Delusion*", is a dismissive and, at times, sarcastic denunciation of belief in God as Creator.[7] The title of the book is indicative of his major theme; that people who believe in God are delusional. Furthermore, he argues that you cannot be an intelligent scientific thinker and still believe in a Creator God. In fact, he states that people who have such a faith are "*infantile*".[8] Yet, in reality, science is not equipped to make any authoritative declarations on the matter of God's existence. The subject simply lies beyond the realm of science's ability to comment.

Sadly, Dr. Dawkins is not the only scientist who ventures beyond the limits of science. For example, Dr. D. M. Watson made the extraordinary statement:

> "*Evolution is a theory universally accepted, not because it can be proved to be true, but because the only alternative, 'special creation', is <u>clearly incredible</u>.*"[9]

The mistake that Drs. Dawkins, Watson and others make, is to assume that because science can only observe and measure things in the natural, physical realm, no other realm exists. That is a completely illogical leap of reasoning. It is one thing to say that science can only measure natural causes for phenomena; it is quite another to insist that no other causes exist. In fact, that is a completely unscientific assumption. Science simply cannot prove that there are no supernatural causes of phenomena in our universe. As Timothy Keller states:

> *"There would be no experimental model for testing the statement: 'No supernatural cause for any natural phenomenon is possible'."*[10]

Science simply cannot say, *"nature is all there is."* That is a completely unprovable proposition. Sadly, however, many atheistic scientists regularly make this very assertion. For example, Sean Carroll, in the previously quoted article, *"Science and Religion are Not Compatible"*, makes the extraordinary comment:

> *"Scientifically speaking, the existence of God is an untenable hypothesis."*[11]

By this, he means that belief in God is *unacceptable* scientifically. He then goes on to say, of scientists who persist in their belief in God:

> *"Those people are just wrong!"*[12]

These are completely unscientific declarations. The existence of God is not an **untenable** hypothesis, it is an **untestable** hypothesis, for it lies outside of science's ability to ultimately prove.

Similarly, just as science cannot say that God does not exist, neither can it say that miracles do not occur. All an individual scientist can say is, *"To date, I have not observed or measured a miracle."* Of course, a miracle would require the temporary suspension or alteration of one or more laws of science, through the intervention of forces beyond the natural realm. Science cannot say, however, that such an intervention and a suspension of the laws cannot happen. If God exists, then the natural laws are merely the normal means by which he sustains the universe. If there is an all-powerful, transcendent God who is underwriting and upholding these natural laws, he is entirely capable of bypassing any one of them, or all of them, momentarily. Certainly, from a faith perspective, the God who is able to create the universe from nothing is more than capable of any and all further interventional creative acts.

Science, therefore, cannot issue authoritative declarations that miracles are not possible, because:

1. It cannot prove a universal negative, and;

2. It cannot comment on the supernatural realm - because science is incapable of observing and measuring it.

It is disappointing, therefore, to read comments such as Sean Carroll's, in his article *"Science and Faith Are Not Compatible"*:

> *"Science and religion are incompatible ... Religion ends up saying things like 'Jesus died and was resurrected' or 'Moses parted the red sea'. And science says none of that is true. So there you go, incompatibility."*[13]

No Sean, science does **not** say *"none of that is true"*. You say that, but science doesn't. In fact, science can't say that. Science has no authority to comment on things beyond its scope of study, nor can it prove a universal negative in that sense.

**The Inability of Science to Travel Back in Time**

Just as science cannot investigate the supernatural realm, it is also not able to observe events in the past. Science is stuck in the present. It can only observe and measure things that exist in the present. Forensic science can observe and measure *artefacts* from the past (such as rocks and fossils), and make *assumptions* about past events based on those artefacts, but those assumptions can never be proven in an absolute scientific sense, because those past events can no longer be observed, measured and repeated in the present. Radiometric dating of rocks is a good example. Science cannot authoritatively tell us how old rocks are. Rocks do not have a date stamp, and apart from rocks formed in recent lava flows, no scientist was alive to witness the formation of rocks in the past. All science can do is observe and measure rocks in the present. They can measure the current composition of rocks and the current decay rates of isotopes, but they have no way of knowing:

- decay rates in the distant past.

- the original composition of rocks and the ratios of parent isotopes to daughter isotopes in the past.

- processes that could have added additional elements to rocks in the past.

- processes that could have removed elements from rocks in the past.

The reason why science cannot know these things is because scientists cannot travel back in time to the past. Thus, when scientists make authoritative declarations about the ages of rocks, they are making huge assumptions about processes in the past for which they have no conclusive evidence. Science faces the same limitations when it comes to fossils. All scientists can do is observe and study fossils in the present. They cannot make authoritative declarations about dates and causes of extinction, because no scientist was there to observe those events.

Forensic science is a sub-genre of science that is dedicated to studying non-observable events of the past, by examining residual evidence in the present. It incorporates estimations and assumptions about past environmental factors and past precipitating factors in order to arrive at a possible explanation and interpretation of the residual evidence in the present. The further back in time the original event took place, the more assumptions and estimates that have to be made, and the less reliable the interpretations become. It is staggering to me, therefore, that evolutionists claim to be able to make authoritative, absolute declarations about events that supposedly took place millions of years ago! It is also remarkable how often scientists have had to modify or even overturn these declarations, as new theories emerge.

**Historical Verification**

The one exception, the one time when scientists *can* make absolute, authoritative statements about the past, is when people *were* present to witness the events, and left behind accurate, verifiable historical records for us to read. I stress the words "accurate" and verifiable", because historical records, before they can be relied upon, must be proven to be accurate. The professionally accepted means by which historical documents are assessed for reliability is referred to as the Criteria of Historical Reliability. I have written about these criteria in detail in my book, "*Finding God When He Seems To Be Hiding*". The criteria include:

- The Criterion of Time Gap

- The Criterion of Embarrassment

- The Criterion of Multiple Attestation

- The Criterion of The Absence of Protestation

- The Criterion of Corroboration by Hostile Critics

- The Criterion of Author Credibility

Once historical accounts have been verified using these criteria, scientists are then, and only then, able to make authoritative declarations about past events. Thus, scientists can state with absolute certainty that Mt. Vesuvius erupted in 79 AD, even though it happened 2,000 years ago and no scientist alive today was there to witness it. This is because the written historical records describing the eruption of Mt. Vesuvius have been verified using the criteria of historical reliability.

It will come as a surprise to many people to learn that the same criteria of historical reliability that verify the eruption of Mt. Vesuvius, also verify the historical accounts of the life, miracles and resurrection of Jesus. Multiple accounts, written by eyewitnesses, within the lifetime of other eyewitnesses, including multiple corroborating accounts by hostile critics apart from the Bible, pass the same criteria of reliability with flying colours. These accounts lead us to conclude that Jesus *did* exist, that he *did* work extraordinary miracles, and that he *did* rise from the dead. In fact there is *much* more written historical evidence for the resurrection of Jesus from the dead than for the eruption of Mt. Vesuvius!

Scientists who accept events such as the eruption of Mt. Vesuvius as an established scientific and historical fact, but dismiss the resurrection and miracles of Jesus as myth, are guilty of filtering the evidence through their preconceived biases. Effectively, they are saying, *"miracles can't happen, because there is no such thing as the supernatural."* As we have already noted, however, such a presupposition is completely unscientific. The existence, or otherwise, of supernatural forces, lies beyond the ability of science to comment. Scientists who make these kinds of declarations are operating from a position of prejudicial personal bias rather than from evidence-based science.

## The Fallibility of Science

Another misconception regarding the nature of science is in regard to its reliability. There is a common tendency among the general populace to regard all scientific declarations as absolute, irrefutable facts. Once a declaration is identified as having come from "science" or "scientists", it is immediately elevated to the status of incontestable truth. Often

## Science and Faith

scientists themselves perpetuate this attitude. For instance, in the Introduction to Jerry A. Coyne's *"Why Science and Religion are Incompatible"*, Neil DeGrasse-Tyson wrote:

> *"The good thing about science is that it's true, whether you believe it or not."*[14]

This is simply not correct. *Facts* are true, whether or not you believe them. Science, on the other hand, has an impressive track record of getting the facts **wrong** and having to retract previously published declarations. Here are some notable examples of scientific "facts", once declared authoritatively to the world, but now known to be false:

- The earth is the centre of the solar system. Contrary to popular belief, it was not just the church that believed this and opposed Nicolaus Copernicus's heliocentric model, but the *whole of the scientific community* who had previously stated unequivocally that the solar system revolved around the earth. The church was simply trusting the weight of previously established scientific opinion.

- Spontaneous generation of life (as previously discussed). Scientists in the middle ages believed and taught that maggots arose spontaneously from rotting meat, frogs from stagnant ponds, earthworms from manure and mice from warm, moist soil.

If you think these kinds of errors were only endemic in medieval science, you are mistaken. Literally dozens of scientific "facts" presented authoritatively to us in recent decades have now been disproven. For example;

- There are nine planets in the solar system. Science now tells us that there are only eight planets, as Pluto has been downgraded.

- The dinosaurs became extinct because of volcanic eruptions. Scientists now believe that asteroid strikes were more likely the cause.

- Neanderthals were pre-humans who died out prior to the advent of modern humans. It is now known that Neanderthals were just a variation of modern humans, living at the same time.

- The "missing links" of human ancestry. All of these have now been discredited. (See Chapter 3, *"The Problem With Fossils"*).

- Part of Einstein's equations introduced the cosmological constant - a force that counteracted gravity and stopped the universe from collapsing in upon itself. He declared that the universe existed in a "steady state", neither expanding nor contracting. This was the prevailing view until the red-shift of light from galaxies, observed by Edwin Hubble and others in 1929, revealed that the universe is actually expanding, as galaxies are moving away from each other. Subsequent to this new discovery, Einstein described the cosmological constant as his biggest mistake, and removed it from his equations.

- Following the Hubble discovery, scientists then made further observations that the universe is *slowing down* in its expansion; expanding more and more slowly, and they announced this to the world. This became an established scientific "fact" for several decades. In the late 1990s, however, this was proved *incorrect* by further remarkable observations from the Hubble Space Telescope and other ground telescopes. Observable Galaxies now appear to be *accelerating* away from each other! Scientists have now amended Einstein's cosmological constant, and reinserted it into his equations.

- The universe has always existed. This was the prevailing view of science from the time of Aristotle, until some remarkable discoveries in the 20th century, which caused the scientific community to completely reverse their declarations. Stephen Hawking commented: *"All the evidence seems to indicate, that the universe has not existed forever, but that it had a beginning. This is probably the most remarkable discovery of modern cosmology."*[15]

These are just a few examples of the many hundreds of instances throughout history where science has "got it wrong" and has had to reverse its declarations. Scientific theories are not "facts"; they are attempts to understand and interpret the facts, and they don't always get it right.

In the light of science's poor track record in regard to truth, it seems extremely presumptuous of evolutionists to insist that people trust science as the ultimate purveyor of truth and to relegate faith to the realm of unreliable myth. This represents an unwarranted elevated view of science's reliability and a complete dismissal of the existence of anything supernatural.

**SCIENTISTS OF FAITH**

Not all scientists are driven by a predetermined antithesis towards faith. In fact, there is a significant percentage of scientists who see no contradiction between their scientific endeavours and their belief in God. A 2009 survey of scientists who are members of the American Association for the Advancement of Science, conducted by the Pew Research Centre, found that 51% of scientists believe in God or some kind of deity or supernatural power.[16] This is a very different picture to the impression given by Richard Dawkins and other outspoken atheists, who maintain that no intelligent scientist could believe in God. Furthermore, even among atheistic scientists, not all of them agree with the prejudicial and disparaging view of faith that Dawkins espouses. Dr. Stephen Jay Gould, the late Harvard evolutionist and atheist, reflecting on the large number of his respected colleagues who had some kind of religious faith, commented:

> *"Either half my colleagues are enormously stupid, or else the science of Darwinism is fully compatible with conventional religious beliefs."*[17]

While I disagree that Darwinism is compatible with the Christian faith, Gould's statement acknowledges the large percentage of scientists who seem to be able to embrace both faith and science without any apparent conflict.

Certainly, this has been the case throughout the ages. In fact, many of the great pioneers of science in the past have been men and women of great faith. Here are some examples, together with their statements of faith:

**Galileo Galilei (1564-1642)**

A pioneer of observational astronomy.

> *"I do not feel obliged to believe that the same God who has endowed us with senses, reason and intellect has intended us to forego their use and by some other means to give us knowledge which we can attain by them. He would not require us to deny sense and reason in physical matters which are set before our eyes and minds by direct experience or necessary demonstrations."*[18]

**Albert Einstein (1879 - 1955)**

Perhaps the greatest scientist to have ever lived, and the physicist who developed the theory of relativity.

> *"I want to know how God created this world. I am not interested in this or that phenomenon, in the spectrum of this or that*

element. I want to know His thoughts; the rest are mere details."[19]

## Lord William Kelvin (1824 - 1907)

Developed the theory of thermodynamics and calculated the value of absolute zero, instigating the still-used Kelvin temperature scale.

> "I believe that the more thoroughly science is studied, the further does it take us from anything comparable to atheism. If you study science deep enough and long enough, it will force you to believe in God."[20]

## Sir Isaac Newton (1643 -1727)

Formulated the laws of motion and universal gravitation.

> "This most beautiful system of the sun, planets, and comets could only proceed from the counsel and dominion of an intelligent Being. This Being governs all things, not as the soul of the world, but as Lord over all; and on account of his dominion he is wont to be called "Lord God" or "Universal Ruler". The Supreme God is a Being eternal, infinite, [and] absolutely perfect."[21]

## Johannes Kepler (1571 - 1630)

Formulated the laws of planetary motion.

> "I wanted to become a theologian. For a long time I was restless. Now, however, behold how through my effort God is being celebrated in astronomy."[22]

## Nicolas Copernicus (1473 -1543)

Discovered and formulated the heliocentric nature of the solar system.

> *"I desire to know the mighty works of God, to comprehend His wisdom and majesty and power; to appreciate, in degree, the wonderful workings of His laws."[23]*

### Louis Pasteur (1822 -1895)

Co-founder of microbiology and immunology, and developed vaccination and pasteurisation.

> *"The more I study nature, the more I stand amazed at the work of the Creator. Science brings men nearer to God."[24]*

### Robert Boyle (1627 - 1691)

The founder of modern chemistry.

> *"God is the author of the universe, and the free establisher of the laws of motion."[25]*

### Max Planck (1858 - 1947)

Theoretical physicist who won the Nobel Prize in Physics in 1918. The originator of quantum theory, and developer of "Planck's Constant", an extremely important constant in cosmology.

> *"God is the beginning and end of all considerations: He is the foundation."[26]*

### Sir Joseph Thomson (1817 - 1901)

Nobel prize winning discoverer of the electron, and the founder of atomic physics.

> *"Every advance in science declares that 'Great are the works of the Lord'."[27]*

## Nikola Tesla (1856 - 1943)

Inventor of electrical alternating current (AC).

> *"The gift of mental power comes from God, and if we concentrate our minds on that truth, we become in tune with this great power."*[28]

## Maria Mitchell (1818 -1889)

America's first female astronomer and the first woman to be entered onto the honour roll of the American Academy of Arts and Sciences.

> *"Scientific investigations, pushed on and on, will reveal new ways in which God works, and bring us deeper revelations of the wholly unknown."*[29]

## CONCLUSION

Faith and science are not contradictory, nor are they mutually exclusive. They deal with two different realms, but those realms are not distinct and separate, for they are inextricably entwined. Life is, at the same time, both physical and spiritual.

Science has its limits though. By its own definition, science can only deal with the observable, measurable natural world. It cannot, therefore, postulate or theorise regarding things that lie outside of that natural world, nor can it claim that nothing outside the natural world exists. This would be akin to a person surrounded by darkness and standing in a pool of light claiming that only those things within the pool of light exist.

If, however, the universe is both physical and spiritual, and if a God exists who is the ultimate cause behind the universe, it would not be unreasonable to expect to observe evidence of his existence and his creative intervention within the physical universe. While ultimate proof or disproof of the existence of a transcendent, supernatural God lies well beyond the remit of true science, this does not rule out the possibility of positive evidence indicating that God's existence is *highly probable*; indeed, almost logically *inescapable*. In fact, Romans 1:20 tells us that God has placed so much evidence for his existence within the natural world, that we are all *"without excuse."*

The overwhelming evidence for God's existence is the topic of the next three chapters.

## ENDNOTES - Chapter 12

[1] Richard Dawkins on militant atheism (transcript), TED talk, 2007 (page visited on 21 January 2015).

[2] Jennifer Wiseman, in an interview with ABC's "Compass" program, quoted in https://www.abc.net.au/news/science/2018-05-24/three-scientists-talk-about-how-their-faith-fits-with-their-work/9543772

[3] http://blogs.discovermagazine.com/cosmicvariance/2009/06/23/science-and-religion-are-not compatible/#.XEll81wzZPY

[4] https://sciencecouncil.org/about-science/our-definition-of-science/

[5] https://www.merriam-webster.com/dictionary/god

[6] Documentary film, "Expelled: No Intelligence Allowed", 2008, directed by Ben Stein.

[7] Richard Dawkins, "The God Delusion", New York, First mariner Books, 2006

[8] Richard Dawkins, "The God Delusion", New York, First mariner Books, 2006, p.360. Also quoted on https://en.wikiquote.org/wiki/Richard_Dawkins

[9] D.M. Watson, London Times, August 3, 1929, cited by Bolton Davidheiser, "Evolution and Christian Faith", Grand Rapids, Baker Book House, 1969, p.79

[10] Timothy Keller, "The Reason For God", Penguin Publishers, 2008, p.86

[11] http://blogs.discovermagazine.com/cosmicvariance/2009/06/23/science-and-religion-are-not-compatible/#.XEII81wzZPY

[12] http://blogs.discovermagazine.com/cosmicvariance/2009/06/23/science-and-religion-are-not-compatible/#.XEII81wzZPY

[13] http://blogs.discovermagazine.com/cosmicvariance/2009/06/23/science-and-religion-are-not-compatible/#.XEII81wzZPY

[14] Neil DeGrasse Tyson, quoted in the Introduction to Jerry A. Coyne, "Why Science and religion are Incompatible", Penguin Books, New York, 2016

[15] Stephen Hawking, "The Beginning of Time", lecture, published on his website, http://www.hawking.org.uk/the-beginning-of-time.html

[16] http://www.pewforum.org/2009/11/05/scientists-and-belief/

[17] Stephen Jay Gould, "Impeaching a Self-Appointed Judge", Scientific American, Vol.267, no.1, 1992, quoted in Alister McGrath, "The Dawkins Delusion?", IVP, 2007, p.34.

[18] Galileo Galilei, in a letter to the Grand Duchess of Tuscany, 1615

[19] From E. Salaman, "A Talk With Einstein," The Listener 54 (1955), pp. 370-371, quoted in Jammer, p. 123

[20] https://www.azquotes.com/quote/587233

[21] Isaac Newton, Principia, Book III; cited in; Newton's Philosophy of Nature: Selections from his writings, p. 42, ed. H.S. Thayer, Hafner Library of Classics, NY, 1953.

[22] Letter to Michael Maestlin (3 Oct 1595). Johannes Kepler Gesammelte Werke (1937-), Vol. 13, letter 23, l. 256-7, p. 40.

23 Nicolas Copernicus, As quoted in Poland : The Knight Among Nations (1907) by Louis E. Van Norman, p. 290; also in The Language of God (2006) by Francis Collins, pp. 230-31

# Science and Faith

[24] Louis Pasteur, quoted in "The Literary Digest", 18 October, 1902.

[25] Robert Boyle (2000). "The Works of Robert Boyle: Publications of 1674, p. 6

[26] Scientific Autobiography and Other Papers as translated by F. Gaynor (1949), p. 184

[27] Joseph Thomson, "Nature" Magazine, Vol. 81, 1909, p.257

[28] Nicola tesla, "My Inventions" first published in Electrical Experimenter magazine (1919); republished as My Inventions : The Autobiography of Nikola Tesla (1983), Chapter 5.

[29] Maria Mitchell, diary entry, 1881.

Chapter 13

# Evidence For A Creator: Cosmology

On 7th April 2008, prominent evolutionist and atheist, Dr. Richard Dawkins, commented in a live podcast interview on *"The Big Questions"*[1]:

> *"There is not a tiny shred of evidence for the existence of any kind of god."*

It was a truly astonishing claim, spoken with the hubris of a scientist who has apparently already made his mind up. This kind of dismissive generalisation is based upon two underlying assumptions:

1. That the natural world is all there is.

2. That everything within our universe can be explained through natural causes.

Both of these assumptions, of course, are completely unprovable. In fact, they are not scientific statements at all; they are purely philosophical beliefs arising directly from an atheistic worldview.

In claiming a complete lack of evidence for the existence of God, Richard Dawkins is out of step with a growing number of cosmologists worldwide who are, in some cases reluctantly, acknowledging the fact that there is considerable scientific evidence pointing towards the existence of some kind of supernatural primary cause. As we shall soon see, many cosmologists have reached the astonishing conclusion that the origin of the universe cannot be explained by natural causes; that it must have been caused *super*naturally.

Just as there is arguably overwhelming evidence contradicting the theory of evolution, so there is equally convincing scientific evidence that our universe was created supernaturally. I use the word *"scientific"* quite deliberately. Evolutionists argue that belief in God can never be considered a scientific worldview, because it ultimately involves belief in something outside of the physical realm which cannot be directly observed, measured or tested by scientific methodology. As was explained in the previous chapter, while this is certainly true of God himself, it is most definitely *not* true of the evidence of his observable, measurable influence within our universe. The physical universe contains extremely convincing scientific evidence of the intervention of an all-powerful, supernatural, intelligent designer who created the natural realm in the beginning and who undergirds the laws that govern its ongoing operation.

Of course, in order to perceive this evidence and evaluate it fairly, we must be willing to examine the facts with an open mind. Therein lies the problem. My own experience in discussing this evidence with sceptics and atheists over many years is that many of them have such a determined disbelief in God, and such a resolute, unshakeable faith that natural processes can explain everything, that they will not even consider the possibility of a supernatural explanation of the evidence. At times, this leads them to cling to the most unlikely and absurd theories in order to avoid conceding the possibility of a much simpler, more obvious spiritual explanation. This intransigent unwillingness to even consider an alternate viewpoint, and to immediately discount anything that does not fit with their own pre-conceived beliefs, is itself unscientific. Yet this is the obstacle that creationists often face when seeking to engage in dialogue with evolutionists; a resolute pre-determination to immediately dismiss anything that does not fit with their theory. As Winston Churchill once said:

> "Men occasionally stumble over the truth, but most of them pick themselves up and hurry off as if nothing had happened."[2]

But for those with an open mind, who are willing to consider the possibility of an alternate explanation for life and who are earnestly searching for the truth, there is very convincing evidence that points to the existence of a supernatural Creator of our universe.

## THE EVIDENCE OF COSMOLOGY

Cosmology is the study of the cosmos - the universe, with its untold billions of galaxies and stars. For the Christian, who looks up in wonder into the night sky, the sheer scale and majesty of the cosmos has always been regarded as a strong evidence for the existence of an almighty Creator. The Psalmist declared:

> "The heavens declare the glory of God; the skies proclaim the work of his hands" (Psalm 19:1)

The response of the atheistic scientist to this cosmological "evidence" has been to simply claim that the universe has always been there. For centuries the Christian has asked the atheist, "*Who made the universe?*", to which the unbelieving scientist has replied, *"No one! It has simply always existed!"* Outspoken atheist, Bertrand Russell (1872 - 1970) once famously stated;

*"The universe is just there, and that's all!"*[3]

This belief in the eternal existence of the universe was the prevailing view of the scientific community from the time of Aristotle (350 B.C.) until the beginning of the 20th century. This completely baseless presupposition was extremely convenient for atheists, because it effectively removed the necessity for believing in any sort of creator. A series of stunning discoveries in the modern era, however, has completely overturned this view of the universe and sent secular scientists scurrying back to their drawing boards.

After the publication of Albert Einstein's theory of general relativity in 1915, Dr. Willem De Sitter published extrapolations of Einstein's theory in 1917, predicting that, if Einstein's theory is correct, the universe should be expanding outwards and, conversely, must have had a finite beginning.[4] This idea of a universe that had a beginning was refuted vigorously by the scientific community at large, including Einstein himself, who wrote to Dr. Sitter, stating, *"This circumstance irritates me"* and in a second letter he stated, *"To admit such possibilities seems senseless."*[5] The scientific community sided with Einstein, and largely ignored Dr. Sitter's theory.

In 1927, cosmologists Dr. Alexander Friedmann, Fr. Georges Lemaitre, Dr. Howard P. Robertson and Dr. Geoffrey Walker published more detailed extrapolations of Einstein's theory, (referred to as the Friedmann–Lemaître–Robertson–Walker metric)[6], confirming Dr. Sitter's findings that the universe must be expanding and lending further weight to the argument that it must have had a beginning.

In 1929, Dr. Edwin Hubble (1889 - 1953), confirmed the expansion of the universe by observing the Doppler redshift of the light from observable galaxies, indicating their recession from earth.[7] In other words, he confirmed from physical observation that the galaxies in the visible cosmos are expanding outwards.

The discovering of cosmic microwave background radiation (CMBR) in 1964 by Drs. Arno Penzias and Robert Wilson[8] (which earned them a Nobel Prize in Physics in 1978) provided the scientific community at that time with what seemed conclusive evidence of residual microwave radiation left over from a "big bang" at the beginning of space-time. Thus, the Big Bang theory of the origin of the universe was formed. This concept that the universe had a beginning, represented a complete reversal of a belief in its *eternal* existence which had been passionately held for over 2,000 years. Dr. Stephen Hawking (1942 - 2018), in a lecture published on his website, commented on this reversal:

> *"All the evidence seems to indicate, that the universe has not existed forever, but that it had a beginning. This is probably the most remarkable discovery of modern cosmology."*[9]

**BIG BANG IN TROUBLE**

In recent decades, some serious doubts have been cast upon the veracity of a "big bang" as an explanation of the universe's creation. One of the major problems is explaining how such a big bang could account for the formation of galaxies and clusters of galaxies, which we observe throughout the universe. If a big bang created the universe, it should have resulted in a fairly uniform dispersal of matter throughout the universe, resulting in a sparsely spread scattering of matter. But this is not what we see. Instead, we find huge clusters of galaxies, densely populated by stars, separated by vast distances of empty space. No computer models of a supposed big bang can show how these clusters could possibly have formed. Dr. James Trefil, professor of physics at Mason University, Virginia, comments:

> "There shouldn't be galaxies out there at all, and even if there are galaxies, they shouldn't be grouped together the way they are.' He later continues: 'The problem of explaining the existence of galaxies has proved to be one of the thorniest in cosmology. By all rights, they just shouldn't be there, yet there they sit. It's hard to convey the depth of the frustration that this simple fact induces among scientists."[10]

Because of these and other observational anomalies that seem to contradict the Big Bang theory of the universe's origin, a growing number of scientists are calling for the theory's official demise. For example, in 1993, cosmologist Dr. Halton C. Arp, of the Mount Wilson Observatory, Pasadena, USA, wrote:

> "In my opinion the observations speak a different language; they call for a different view of the universe. I believe that the big bang theory should be replaced, because it is no longer a valid theory."[11]

The expansion of the universe has also been challenged recently by some observational data which suggests alternate explanations for redshift in galaxies and for cosmic microwave background radiation (CMBR).[12]

These and other recent developments have led a growing number of respected scientists to reject the notion of a big bang. In 2004, an "Open Letter To The Scientific Community" by 33 leading scientists has been published on the internet,[13] and republished in *"New Scientist"*.[14] A subsequent article was published on www.rense.com, entitled *"Big Bang Theory Busted by 33 Top Scientists"*.[15]

Thus, modern cosmology is a long way from consensus on these issues. Despite the divergence of opinions however, there is very clearly a groundswell of scientists willing to concede the existence of a creative force or forces instrumental in the universe's formation. Even those who continue to hold to the big bang theory as the cause of the universe's beginning cannot adequately explain the observable complexity of the universe. Something else had to have been at work, shaping the universe at the beginning of time.

## THE UNIVERSE HAD A BEGINNING

Despite controversy and growing scientific scepticism regarding the big bang as an *explanation* for the universe's beginning, most scientists agree that the evidence of modern physics and cosmology points to the inexorable conclusion that the universe began *somehow*. The process of that beginning is unclear, but its fact is now almost universally accepted. This is further confirmed by several other very powerful theoretical arguments.

### The Second Law of Thermodynamics

Newton's second law of thermodynamics indicates that in any closed system, the total amount of energy available diminishes over time. This is referred to as entropy, and, when applied to cosmology, it simply means that the universe is gradually winding down. In turn, this means that over time the universe is getting progressively colder. Our observation of the nature and function of stars confirms this. Because stars are finite bodies with finite mass and energy it is not possible for them to burn continuously for eternity. Given sufficient time, each star will eventually burn itself out of existence. This is a simple logical extension of Newton's theory. This, in itself, provides conclusive proof that the universe cannot have existed forever. If the universe has existed for eternity past, it would have burnt itself out by now!

### The Philosophical and Mathematical Impossibility of an Eternal Universe

Al-Ghazali, a Persian philosopher in the Middle Ages, proposed this simple philosophical proof that the universe cannot have existed forever. He stated:

> *"If there were an infinite number of events in the past, we would never have arrived at the present"*[16]

This simple, yet profound philosophical argument cannot be refuted without dismissing the laws of logic. Think about it. If time is a corridor and you started in the present and went back in time towards infinity past, you would never reach infinity past. You can't reach infinity past from the present. You will never get there! In the same way, you can't reach the present from infinity past. It's impossible! Al-Ghazali's theorem proves that the universe cannot have existed forever.

Similarly, Al-Ghazali also proposed that:

> *"An infinite number of things cannot exist."*[17]

This is because infinity cannot logically exist in a physical universe. David Hilbert was a German Mathematician who came up with a brilliant way of explaining this concept. He asks us to imagine a hotel with an infinite number of rooms, each occupied by a guest. Therefore, the number of guests is infinity. Suppose one more guest arrives and wants a room. The hotel manager asks every guest to move to the next room number. The guest in room 1 moves to room 2. The guest in room 2 moves to room 3. The guest in room 5,000,000,000 moves to room 5,000,000,001 and so on. Now there is a vacant room in Room 1, and the new guest is accommodated. But the number of guests still numbers infinity. Thus:

**Infinity + 1 = Infinity**

This works for the addition of any finite number. It also works for subtraction.

But suppose an infinite number of new guests arrive, each requiring a room. No problem! Every existing guest is simply asked to move to a room number twice their existing number. Thus, Room 1 moves to room 2. Room 2 moves to room 4. Room 3 moves to room 6, and so on. In this way the pre-existing infinity of guests are now all in even numbered rooms, leaving an infinity of odd numbered rooms vacant. The infinity of *new* guests can now move into the odd numbered rooms.

But how many guests are there now? Still infinity! Thus:

### Infinity + infinity = infinity

But suppose everyone in the even rooms now leave? An infinite number of guests were in the hotel. An infinite number of guests now leave (even numbers) but an infinite number of guests still remain (odd numbers). Thus,

### Infinity - infinity = infinity

What on earth has all this got to do with the creation /evolution debate? Quite simply, it shows how absurd infinity is as an actual number in a physical universe. Infinity exists only as an abstract concept; it cannot exist as an actual number in a physical universe. Therefore, there cannot have been an infinite number of past events and past days, hours or seconds. Therefore, the universe must have had a beginning!

The evidence of cosmology, philosophy and mathematics all point to a clear, logical conclusion: The universe has not existed forever. It had a beginning. As Dr. Stephen Hawking stated on his website:

> *"We have made tremendous progress in cosmology in the last hundred years ... which has shattered the old picture of an ever-existing and ever-lasting universe. ... This is a profound change in our picture of the universe and of reality itself."*

**THE NECESSITY OF A SUPERNATURAL CAUSE**

This brings us to a final, inescapable, astounding, logical conclusion; something *beyond* the physical universe had to have created it.

# Evidence For A Creator: Cosmology

Most cosmologists are now reaching the astonishing conclusion that at some time in the distant past there was nothing at all. No stars, no planets, no asteroids, no gases, no chemicals, no elements, no physical matter at all. Nothing. And then a moment later, there was a universe! Stephen Hawking was correct in saying that this is *"a profound change in our picture of the universe and of reality itself."* Because if the universe had a beginning, it raises the question: Who put it there? How can **nothing** become **something** unless **someone** beyond the universe creates it? The cosmological evidence for the universe having a beginning cries out for a supernatural explanation, because only something **outside** of nature, something "supernatural", can possibly create nature. Nature cannot create itself out of nothing!

## The Kalam Cosmological Argument

The Kalam cosmological argument deals with the issue of ultimate cause. It is a philosophical argument, originating in the Middle Ages, and championed in recent years by apologist and theologian, Dr. William Lane Craig. The original Kalam argument is as follows:

1. Whatever begins to exist has a cause.
2. The universe began to exist.
3. Therefore, the universe has a cause.

William Lane Craig added to the argument:

3. That cause must be timeless, immaterial, self-existent and powerful.
4. The most reasonable cause is God.

This is a logical argument that, in my opinion, is impossible to refute. The universe can't have created itself, therefore the cause must lie outside of the universe! Of course, there will always remain a core of obdurate atheists who refuse to concede even this clear chain of logic. For example, Dr. Quentin Smith, professor emeritus of philosophy at Western Michigan University, in a debate with William lane Craig, stated:

> "The universe came from nothing, by nothing for nothing!"[18]

This, of course is a ludicrous proposition, and it illustrates that some atheists are prepared to believe in the impossible, rather than the supernatural.

A growing number of scientists, however, are concluding that there was something beyond nature that was fundamentally at work in the creation of the universe. In April 2016, Dr. Dan Reynolds wrote:

> "All the observable evidence we have about the universe implies it had a beginning ... Logically, the universe did not and could not create itself. If the universe (nature) could/did not create itself and it had a beginning, then only something or someone outside of nature can account for the universe's existence. Genesis 1:1 offers a credible explanation: In the beginning God created the heaven and the earth."[19]

Dr Robert Jastrow, astronomer, physicist and founder of NASA's Goddard Institute of Space Studies, stated,

> "Astronomers now find they have painted themselves into a corner because they have proven, by their own methods, that the world began abruptly in an act of creation to which you can trace the seeds of every star, every planet, every living thing in this cosmos and on the earth. And they have found that all this happened as a product of forces they cannot hope to discover. That there are what I or anyone would call supernatural forces at work is now, I think, a scientifically proven fact."[20]

Similarly, Dr James Clerk Maxwell, physicist and mathematician, who is credited with formulating classical electromagnetic theory and whose contributions to science are considered to be of the same magnitude to those of Einstein and Newton, stated:

> *"Science is incompetent to reason upon the creation of matter itself out of nothing. We have reached the utmost limit of our thinking faculties when we have admitted that because matter cannot be eternal and self-existent it must have been created."*[21]

Commenting on the growing number of scientists who now concede that the universe must have had a supernatural cause, astrophysicist, Dr Hugh Ross, Director emeritus of Observations at Royal Astronomical Society, Vancouver, states;

> *"Astronomers who do not draw theistic or deistic conclusions are becoming rare, and even the few dissenters hint that the tide is against them. Geoffrey Burbidge, of the University of California at San Diego, complains that his fellow astronomers are rushing off to join 'the First Church of Christ of the Big Bang.'"*[22]

In other words, there is a growing tide of scientists at the top of their fields who, when confronted with the mounting cosmological evidence, are conceding that the only logical explanation for the origin of the universe, is that there must have been a supernatural cause. Nature cannot create itself; therefore the cause had to have been something outside of nature - a transcendent *"supernatural"* cause. This does not mean that all the scientists in the world are suddenly becoming Christians. There is a big step from believing in a supernatural cause of some kind, to believing in the God of the Bible. But for the first time in a long time, a growing chorus of voices within the scientific community has conceded the very real possibility of the existence of supernatural forces that lie beyond the realm of scientific study.

Dr Hugh Ross states;

> *"All the data accumulated in the twentieth and twenty-first centuries tell us that a transcendent Creator must exist. For all*

> the matter, energy, nine space dimensions, and even time, each suddenly and simultaneously came into being from some source beyond itself. Likewise, it is valid to refer to the Creator as transcendent, for the act of causing these effects must take place outside or independent of them."[23]

Of course, this is not the impression that continues to be dished up to us by the media. The humanist movement has a stranglehold on the popular press. Scientists with aggressive atheistic agendas dominate the airwaves, and their well-financed documentaries continue to pump out the message that science has a natural explanation for everything. The impression is given that the new god of science has all the answers sewn up. But scientists who are at the cutting edge of their fields know that this is not so. Those who are studying these things in depth are coming face-to-face with undeniable evidence of a transcendent, supernatural reality that underpins our entire universe. The tide is turning, and it's not just a trickle. The previously quoted complaint of Dr. Geoffrey Burbidge, of the University of California at San Diego, that his fellow astronomers are *"rushing off to join the First Church of Christ of the Big Bang"*[24] indicates how widespread this new spiritual awareness is within the scientific community.

Without a doubt, the science of cosmology provides extremely convincing evidence for the existence of a supernatural creator-God. The extraordinary claim by Richard Dawkins, that *"there is not a tiny shred of evidence for the existence of any kind of god"*[25], must surely arise from a wilful determination to ignore the considerable cosmological evidence that has arisen in recent years, and portrays his intransigent unwillingness to even consider the existence of anything beyond the realms of science.

## THE DESPERATE SEARCH FOR AN ALTERNATE EXPLANATION

As the evidence of cosmology increasingly points to a supernatural cause, die-hard atheists are becoming ever more desperate in their search for an alternate explanation for the origin of the universe that does not involve a personal God. These alternate explanations include:

- **Bubble universes:**[26] Also known as "eternal inflation theory". An endless series of new bubble universes, expanding and breaking off from existing universes, like giant boils. (But where did the first universes come from?)

- **An oscillating or cyclical universe:**[27] The idea that our universe is endlessly expanding then contracting back to a single point again, then exploding in a big bang again, on and on forever. (But logic and the second law of thermodynamics have both proved that a physical universe cannot be eternal!)

- **Baby universes:**[28] One of Stephen Hawking's desperate postulations theorised that black holes are the umbilical cords that connect baby universes with the mother universes that gave birth to them. (But where did the mother universes come from?)

- **Aliens created our universe:**[29] Proposed by Richard Dawkins and others. (Oh dear! Really? In that case, who created the aliens???)

- **Our future selves created the universe:**[30] Our future selves eventually developed time travel and travelled back to the distant past and created the physical universe for our past selves to evolve into. (I'm not making this up! This has been suggested in scientific circles!)

- **Simulation theory:**[31] Our universe is simply a giant computer simulation designed by aliens or, once again, our super-advanced future selves, and we are all simply plugged into the simulation. This theory was proposed by Dr. Nick Bostrom, professor of Physics at Oxford University in 2003.[32] ("The Matrix" is apparently true!)

- **Something from nothing:** In another of Stephen Hawking's desperate attempts to avoid the existence of a creator-God, he postulated, *"Because there is a law such as gravity, the universe can and will create itself from nothing."*[33] (This effectively throws all the laws of science and the accepted laws of cause and effect, out the window!)

- **We don't know:** Those less inclined to scientific fairy stories, but still equally dismissive of the supernatural, simply state that we don't know how the universe was initially formed. An article on the website, American Scientist, commented: *"The question of how matter came into existence in the formation of the universe still awaits a satisfactory answer."*[34] (At least this is honest!)

Christopher J. Isham, Britain's leading quantum cosmologist, and an astrophysicist at Imperial College of London, recently wrote:

> *"The idea that the Big Bang supports theism is greeted with obvious unease by atheist physicists. At times this has led to <u>wild scientific theories</u> being advanced with a tenacity which so exceeds their intrinsic worth that one can only suspect the operation of psychological forces lying very much deeper than the usual academic desire of a theorist to support his or her theory."*

Dr. Isham is absolutely correct. These wild and bizarre theories indicate a desperate determination to avoid belief in a supernatural Creator, and a corresponding willingness to embrace anything, no matter how fanciful, in order to do so. As a Christian, this does not surprise me in the least. As the Apostle Paul once wrote:

#  Evidence For A Creator: Cosmology

*"The god of this age has blinded the minds of unbelievers, so that they cannot see the light of the gospel that displays the glory of Christ, who is the image of God." (2 Cor 4:4)*

---

## ENDNOTES - Chapter 13
---

[1] Richard Dawkins interview, https://www.youtube.com/watch?v=of-8Q3HySjE&t=44m08s

[2] https://www.brainyquote.com/quotes/winston_churchill_135270

[3] Debate between Bertrand Russell and Fr. F. C. Copleston, broadcast on BBC radio, 1948, transcript: http://www.biblicalcatholic.com/apologetics/p20.htm

[4] https://history.aip.org/exhibits/cosmology/ideas/expanding.htm

[5] https://history.aip.org/exhibits/cosmology/ideas/expanding.htm

[6] https://en.wikipedia.org/wiki/Friedmann%E2%80%93Lema%C3%AEtre%E2%80%93Robertson%E2%80%93Walker_metric

[7] https://en.wikipedia.org/wiki/Hubble%27s_law

[8] https://en.wikipedia.org/wiki/Cosmic_microwave_background

[9] http://www.hawking.org.uk/the-beginning-of-time.html

[10] J. Trefil, *The Dark Side of the Universe.* Charles Scribner's Sons, Macmillan Publishing Company, New York, USA, pp. 3, 55, 1988.

[11] Halton C. Arp, quoted in E.P. Fischer (Ed.), *Neue Horizonte 92/93—Ein Forum der Naturwissenschaften—Piper-Verlag*, München, Germany, pp. 112–173, 1993

[12] https://creation.com/expanding-universe-2

[13] http://blog.lege.net/cosmology/cosmologystatement_org.pdf

[14] Lerner, E., Bucking the big bang, *New Scientist* **182**(2448)20, 22 May 2004

[15] "Big Bang Theory Busted by 33 Top Scientists", www.rense.com, 27 May 2004.

[16] Al-Ghazali, quoted in Greg dewar, "Advanced Philosophy and Ethics of Religion", Oxford University Press, Oxford. 2002, p.18.

[17] Al-Ghazali, quoted in Greg dewar, "Advanced Philosophy and Ethics of Religion", Oxford University Press, Oxford. 2002, p.18.

[18] Debate between William Lane Craig and Quentin Smith, https://www.reasonablefaith.org/media/debates/does-god-exist-the-craig-smith-debate-2003/

[19] https://tasc-creationscience.org/article/scientific-evidence-points-creator

[20] https://en.wikipedia.org/wiki/Robert_Jastrow

[21] James Clerk Maxwell; Perspectives on His Life and Work", Oxford University Press, 2014, p.274

[22] Dr Hugh Ross, "The Creator and The Cosmos", Navpress, 2001, pp.108-112).

[23] Dr Hugh Ross, "The Creator and The Cosmos", Navpress, 2001, pp.108-112).

[24] Quoted by Hugh Ross, "The Creator and The Cosmos", Navpress, 2001, pp.108-112

[25] Richard Dawkins interview, https://www.youtube.com/watch?v=of-8Q3HySjE&t=44m08s

[26] Proposed by Paul Steinhardt and Alexander Vilenkin, in 1983. https://en.wikipedia.org/wiki/Eternal_inflation

[27] https://en.wikipedia.org/wiki/Cyclic_model

[28] Stephen Hawking, "Black Holes and Baby Universes", Random House Publishers, 1994.

[29] Video Clip of interview between Ben Stein and Dr. Richard Dawkins, "Richard Dawkins Believes Extraterrestrials Created Man." https://www.youtube.com/watch?v=AiVoS78lNqM

[30] https://www.abc.net.au/news/science/2018-09-02/block-universe-theory-time-past-present-future-travel/10178386

[31] https://www.theguardian.com/technology/2016/oct/11/simulated-world-elon-musk-the-matrix

[32] https://www.theguardian.com/technology/2016/oct/11/simulated-world-elon-musk-the-matrix

[33] Stephen Hawking, cited in *Michael Holden (2010-09-02)*. "God did not create the universe, says Hawking". *Reuters*. Retrieved 2010-10-17

[34] Article on Americanscientist.org, March 2017, no longer available.

Chapter 14

# Evidence For A Creator:
# Intelligent Design

The Apostle Paul, in his opening chapter of his masterpiece, the letter to the Romans, writes this:

> "The wrath of God is being revealed from heaven against all the godlessness and wickedness of people, who suppress the truth by

> *their wickedness, since what may be known about God is plain to them, because God has made it plain to them. For since the creation of the world God's invisible qualities - his eternal power and divine nature - have been clearly seen, being understood from what has been made, so that people are without excuse."* (Romans 1:18-20).

To paraphrase, Paul is saying that God has made his existence abundantly obvious by leaving clear evidence of his greatness in the very fabric of the universe itself. The physical world contains within it, clear signs of an intelligent designer.

The concept of intelligent design has received much attention in recent years. As scientists have grown in their understanding of the complex forces and structures undergirding our universe, it has become increasingly less likely that the universe is the product of brute chance. Even among those scientists who refuse to believe in a personal God, many are having to concede that some form of intelligence was at work in the design and formation of these fundamental forces and structures.

Walter Bradley, a retired professor of mechanical engineering at Texas A&M University, a towering figure within the intelligent design movement, and co-author of "The Mystery of Life's Origin", stated:

> *"It is quite easy to understand why so many scientists have changed their minds in the past thirty years, agreeing that the universe cannot reasonably be explained as a cosmic accident. Evidence for an intelligent designer becomes more compelling the more we understand about our carefully crafted habitat."*[1]

Evidence of intelligent design is everywhere, but in this chapter, we will limit our discussion to a few key examples.

# Evidence For A Creator: Intelligent Design

## INTELLIGENT DESIGN IN PHYSICS

The fundamental forces that underpin our universe are many and complex. These are the forces that hold the very fabric of our universe together, without which, matter itself could not exist. These forces are called universal or physical "constants", because they do not change, and they are incredibly fine tuned to be precisely what is necessary for a life-permitting universe to exist. There are 34 recognised universal constants, from the tiny elementary charge that holds atoms together to the cosmological constant that keeps the universe from flying apart or from collapsing in upon itself.

### Table of Universal Constants

| Quantity | Symbol | Numerical value | Unit |
|---|---|---|---|
| Acceleration of free fall (standard) | $g_n$ | 9.8066 | $m/s^2$ |
| Atmospheric pressure (standard) | $p_0$ | $1.0132 \times 10^5$ | Pa |
| Atomic mass unit | $u$ | $1.6606 \times 10^{-27}$ | kg |
| Avogadro constant | $N_A$ | $6.0220 \times 10^{23}$ | $mol^{-1}$ |
| Bohr magneton | $\mu_B$ | $9.2741 \times 10^{-24}$ | $J/T$, $A\,m^2$ |
| Boltzmann constant | $k$ | $1.3807 \times 10^{-23}$ | $J/K$ |
| Electron | | | |
|   charge | $-e$ | $1.6022 \times 10^{-19}$ | C |
|   mass | $m_e$ | $9.1095 \times 10^{-31}$ | kg |
|   charge/mass ratio | $e/m_e$ | $1.7588 \times 10^{11}$ | $C/kg$ |
| Faraday constant | $F$ | $9.6485 \times 10^4$ | $C/mol$ |
| Free space | | | |
|   electric constant | $\varepsilon_0$ | $8.8542 \times 10^{-12}$ | $F/m$ |
|   intrinsic impedance | $Z_0$ | 376.7 | $\Omega$ |
|   magnetic constant | $\mu_0$ | $4\pi \times 10^{-7}$ | $H/m$ |
|   speed of electromagnetic waves | $c$ | $2.9979 \times 10^8$ | $m/s$ |
| Gravitational constant | $G$ | $6.6732 \times 10^{-11}$ | $Nm^2/kg^2$ |
| Ideal molar gas constant | $R$ | 8.3144 | $J/(mol\,K)$ |
| Molar volume at s.t.p. | $V_m$ | $2.2414 \times 10^{-2}$ | $m^3/mol$ |
| Neutron rest mass | $m_n$ | $1.6748 \times 10^{-27}$ | kg |
| Planck constant | $h$ | $6.6262 \times 10^{-34}$ | $J\,s$ |
|   normalised | $h/2\pi$ | $1.0546 \times 10^{-34}$ | $J\,s$ |
| Proton | | | |
|   charge | $+e$ | $1.6022 \times 10^{-19}$ | C |
|   rest mass | $m_p$ | $1.6726 \times 10^{-27}$ | kg |
|   charge/mass ratio | $e/m_p$ | $0.9579 \times 10^8$ | $C/kg$ |
| Radiation constants | $c_1$ | $3.7418 \times 10^{-16}$ | $W\,m^2$ |
| | $c_2$ | $1.4388 \times 10^{-2}$ | $m\,K$ |
| Rydberg constant | $R_H$ | $1.0968 \times 10^7$ | $m^{-1}$ |
| Stefan-Boltzmann constant | $\sigma$ | $5.6703 \times 10^{-8}$ | $J/(m^2\,K^4)$ |
| Wien constant | $k_w$ | $2.8978 \times 10^{-3}$ | $m\,K$ |

These recognised universal constants are so finely tuned that altering one by even the tiniest fraction of one percent would render life and, in the case of some of the constants, the existence of matter itself, impossible. Take, for example, the elementary charge, also known as the electron charge or the strong nucleic force. This is the attractive force that exists between electrons and protons within an atom, which has been determined to be 1.602176634×10–19 C (coulombs). This tiny force is precisely what is required to keep electrons in stable orbit around the nucleus of an atom. If the elementary charge was even one billionth of one percent greater, electrons would collapse into the nucleus of the atom. If the charge was only one billionth of one percent weaker, electrons would no longer be held in stable orbit and would fly away from the nucleus. In either case matter, as we know it, would cease to exist.

Gravity is another example. This the attractive force that matter exerts upon matter. It is the force that holds us to the surface of the earth. It is the force that holds the very earth together. Gravity is the force that allows planets and stars to form and remain a stable mass. It is a fundamental force that holds the fabric of our universe together. The force of gravity has been measured to be $6.6732 \times 10^{-11}$ N m$^2$ kg$^{-2}$. This is a surprisingly small force. Let me expand it out so you can see how really small it is: 0.000000000066732 N m$^2$ kg$^{-2}$. This is why, when you sit next to someone, you cannot feel the gravitational attraction between you and the other person (although, depending on the other person, you may feel a different kind of attraction!). Gravity only becomes noticeable when at least one of the objects in the equation is very massive - like a planet. The gravitational constant is extremely finely tuned. Dr. Robin Collins, professor emeritus of theoretical physics at NorthWestern University, states:

> *"If the gravitational constant was increased by just one part in ten thousand billion billion billion, that small adjustment would increase gravity by a billion-fold!"*[2]

In such a case all matter would collapse in upon itself and the universe would consist of nothing but black holes. Conversely, if the gravitational constant was decreased by a similarly minute amount, matter would simply drift apart. There would be no stars, no planets, no basis for physical life.

The so-called weak nucleic force is another physical constant, which operates within the nucleus of an atom. It is so finely tuned that altering its value by even one part in $10^{100}$ would prevent the very existence of matter. Similarly, the cosmological constant, the energy density of "empty" space, is even more precisely fine tuned than gravity. It needs to be precisely at its current value in order for the universe to exist at all. Dr. William Lane Craig states:

> *"A change in the value of the so-called cosmological constant, which drives the acceleration of the universe's expansion, by as little as one part in $10^{120}$ would have rendered the universe life-prohibiting."[3]*

To give you an idea of how staggeringly huge these numbers are, and therefore, how extraordinarily narrow is the degree of this fine tuning, consider the following numbers. If the universe is 14 billion years old (as evolutionists suggest), this equates to $10^{17}$ seconds (1 followed by 17 zeros). The number of atoms in the entire universe has been estimated to be $10^{80}$. These are incomprehensibly huge numbers.[4] And yet the fine tuning of many of the universal constants involve precision down to one part in quadrillions of times larger numbers than these!

This incredible fine tuning is the same with all the physical constants. Each of them is precisely tuned to an infinitesimally exact strength that is essential for the universe to exist and for life to be possible. To change any of them by even the infinitesimally smallest of fractions would render the universe as we know it no longer possible.

The possibility of these universal constants all arriving at their precise values by sheer chance alone is so infinitesimally small as to be an impossibility. Given that each of these forces could have been formed at *any* possible value, from extremely strong to extremely weak, the chance of them *all* arriving at precisely the exact strength necessary for a life-permitting universe is simply astonishing! Professor of Philosophy, Dr Robin Collins states:

"The chance of just two of these cosmological constants developing by sheer chance, is one in 100 million trillion trillion trillion trillion trillion. That's more than the number of atoms in the universe! And that's just TWO of the constants!". [5]

But that refers to just two of the constants. The probability of all 34 constants spontaneously appearing at their precise strengths simultaneously is calculated to be about 1 in $10^{600}$ (1 with 600 zeros after it)! To give you an idea of how impossibly large $10^{600}$ is, the number of atoms in the entire universe is calculated to be $10^{80}$. So the chance of just two constants "evolving" spontaneously by chance, is 1 chance in $10^{520}$ times the number of atoms in the universe! That is the number of atoms in our universe multiplied by $10^{520}$ further universes! In other words, it is a complete statistical impossibility that all these precisely tuned constants could have evolved by chance.

Dr Paul Davies, Professor of Theoretical Physics, Adelaide University, states, *"The physical universe is put together with an ingenuity that is so astonishing, with physical constants that are so impossibly perfect, that I can no longer accept it as the product of brute chance".* [6]

Dr. Fred Hoyle, astrophysicist and mathematician, Cambridge University, states, *"A common sense interpretation of the facts suggests that a super-intellect has monkeyed with physics, as well as with chemistry and biology, and that there are no blind forces worth speaking about in nature. The numbers one calculates from the facts seem to me so overwhelming as to put this conclusion almost beyond question."*

Albert Einstein, in a letter to a friend, Phyllis, on January 24, 1936, wrote:

> *"Everyone who is seriously involved in the pursuit of science becomes convinced that some spirit is manifest in the laws of the*

*universe, one that is vastly superior to that of man. In this way the pursuit of science leads to a religious feeling of a special sort, which is surely quite different from the religiosity of someone more naïve."*

## THE EVIDENCE OF ASTRONOMY

Similar to the fine-tuning of physics, the position of the earth within the solar system and within the galaxy has been specifically designed to offer the perfect conditions for biological life. The earth is the perfect distance from the sun. It orbits the sun in a nearly perfectly circular orbit in what is referred to as the circumstellar habitable zone, or the "Goldilocks zone" (it is "just right" for biological life). A little closer, and all water would evaporate, and a little further away and we would be a frozen planet. The earth's size, mass, rate of spin and axis of tilt (23.5 degrees) are also absolutely perfect for biological life to survive and flourish. Our atmosphere has the perfect combination of gases necessary for life. The earth's magnetic field is precisely the right strength, acting as a shield, protecting us from the majority of harmful radiation from the sun. The moon also plays a vital role in creating the tidal forces necessary for circulating and oxygenating the earth's water systems. At around one-eightieth the mass of the earth, the moon is exceptionally large for the size of the earth when compared to the 60 moons of other planets in our solar system. This ratio of moon to planet is unique among all the planets with moons that have been discovered in our galaxy, yet this is the perfect size for the life-giving function that it fulfils for our planet. The sun, too, is just the right kind of star; a yellow dwarf, main sequence star. It is the right size, the right temperature, the right spectral class and the right mass. Only about 7% of stars in the observable universe would be right for us.

The location of our solar system within the Milky Way galaxy is also remarkably ideal. If we were closer to the galactic core, the huge amount of radiation would render life as we know it impossible. Our location on the outer edge of a spiral arm of the galaxy keeps us at a safe distance from the maelstrom of deadly radiation at the galactic core. Propitiously, it also places us in an ideal position to observe the universe. If our solar system was embedded more deeply in the densely populated galactic core, all we would see when we looked up into the night sky would be a thick, impenetrable carpet of stars. But our position on the outer edge of the galaxy allows us to not only observe the shape and nature of our own galaxy from a wonderful vantage point (side on), but also enables us to look away from our own galaxy, in the opposite direction, and observe billions of other galaxies in the rest of the universe. It is as if our planet was deliberately placed in the perfect position to not only keep us safe, but also to enable us to discover the universe for ourselves.

Another fascinating "coincidence" is our ability to observe solar eclipses. These occur when the moon passes directly in front of the sun, momentarily eclipsing it. These solar eclipses have enabled scientists to study the sun's corona (outer atmosphere) in a way that would otherwise be impossible. Solar eclipses have also led to some startling scientific discoveries regarding the nature of the sun and the properties of light, as scientists have been able to observe the refraction of light around the sun. In fact, it is through observations of solar eclipses, that scientists have made many important discoveries about the nature of our universe. For example, during an eclipse, we are able to glimpse stars that are behind the sun, as the sun's gravity bends the light from those stars around the sun itself, making them visible to us. By observing the ability of gravity to bend light in this way, scientists were able to confirm Einstein's general theory of relativity.

What makes solar eclipses so special, is how precisely the position and size of the earth, moon and sun need to be in order for an eclipse to be possible. What is extraordinary is that, when viewed from the perspective of earth, the moon appears to be **exactly** the same size as the sun, so that when it passes in front of the sun it **exactly** covers the sun, while leaving the sun's outer atmosphere visible. For about two minutes during an eclipse, the moon covers the intensely bright photosphere of the sun, enabling us to observe its thin faint chromosphere and the spectacular corona with its dramatic prominences. This is because the sun is 400 times bigger than the moon, but it is also 400 times further away. In fact, the ratio of the size of the moon when compared to the size of the sun is **exactly the same ratio** as their distances from earth! This is an extraordinary "coincidence" which baffles atheistic scientists. The chances of this astronomical arrangement arising by sheer chance is too astonishingly small to be considered possible. It is as if "Someone" deliberately designed our earth and our solar system to enable us to conduct scientific investigations into the nature of the universe.

Of course, evolutionists love to talk up the probability of the existence of earth-like, habitable exoplanets (planets outside our solar system). They would love to prove that life is not as miraculous as creationists propose. In recent years there have been frequent "announcements" from astronomers estimating that there may be anywhere from 10 million to 100 billion habitable earth-like exoplanets in the universe. Significantly, of the several thousand exoplanets discovered to date, none of them are even remotely earth-like or habitable. The Planetary Habitability Laboratory website is run by the University of Puerto Rico in conjunction with data from the Keppler space telescope. It keeps an up-to-date list of potentially habitable planets. Currently they list ten exoplanets that "may" be the right distance from their sun. The only one remotely similar in size to earth, Proxima Centauri B, is twice the size of earth, is tidally locked to its sun (one face permanently facing the sun) and orbiting a red dwarf star (the wrong kind of star). Wikipedia's entry for Proxima Centauri B, states:

> *"Its habitability has not been established, though it is unlikely to be habitable since the planet is subject to stellar wind pressures of more than 2,000 times those experienced by Earth from the solar wind."[7]*

What an extraordinary statement! *"Its habitability has not been established"*? Really? I would say its **lack** of habitability has been firmly established.

The more exoplanets we discover, the more unique our Earth appears to be. Recently, astrophysicist Dr. Erik Zackrisson, an astrophysicist at Uppsala University in Sweden, published estimates on the likelihood of finding an earth-like planet in the universe. He designed a computer modelling program which took into account all the extremely precise conditions essential for a life-producing earth-like planet, and factored in the universal constants and laws of physics. The probability of an earth-like planet being produced by chance processes came out as 1 in 700 quintillion (7 with 20 zeros after it).[8] Given our current estimates of the number of planets in the entire universe, Dr. Zackrisson concluded that the Earth is probably the only planet of its kind in the entire universe.

In 2015, John Horgan, a deist rather than a Christian, wrote an article in *"American Scientific"*, entitled *"Can Faith and Science Co-exist?"*. In it, he said:

> *"The more scientists investigate our universe, the more improbable our existence seems. If you define a miracle as an infinitely improbable event, then <u>our existence, you might say, is a miracle</u>. Scientists try in vain to hand-wave our improbability away ... My own mystical intuitions keep me from ruling out the possibility of supernatural creation."[9]*

Evidence For A Creator: Intelligent Design

**THE EVIDENCE OF BIOCHEMISTRY & GENETICS**

In previous chapters we have already delved into the incredibly complex micro-world of cells and DNA. I will not duplicate that information again here in detail, but simply reiterate the main elements that point to an intelligent designer:

- DNA: The incredibly complex coding built into almost every living cell, with 3.2 billion pieces of information in the human genome. This biological programming is literally millions of times more complex and sophisticated than any computer software ever written, and there is no possible natural means by which this vast reservoir of information could have come into existence by natural means. DNA cries out for an intelligent designer.

- The irreducible complexity of the single cell: Hundreds and thousands of interdependent molecular machines all carrying out their specific functions within the cell, making it a highly organised, complex biological factory. In the case of many of these molecular machines, if they were not present and functional, the cell would not be a living organism. Random natural processes, even those supposedly driven by natural selection, could never have produced such a complex biological factory. It is overwhelming evidence for an intelligent designer.

- The biochemical origin of life itself: Non-living matter could never have produced a living organism in the beginning. Abiogenesis (the creation of life from non-living matter) by random natural means is a fantasy. The original creation of life from non-living matter unequivocally reveals the hand of an intelligent designer.

These amazing biological systems, which far exceed the capacity of human technology to mimic or even fully understand, point toward a transcendent Creator. As renowned biochemist, Dr. Michael Behe, states:

> *"My conclusion can be summed up in a single word: design. I say that based on science. I believe that irreducibly complex systems are strong evidence of a purposeful, intentional design by an intelligent agent."*[10]

Sir Isaac Newton once reportedly said:

> *"In the absence of any other proof, the thumb alone would convince me of God's existence."*[11]

Biologist Dr. Stephen Meyer, in an interview with journalist and author Lee Strobel, stated:

> *"Information is the hallmark of a mind. And purely from the evidence of genetics and biology, we can infer the existence of a mind that's far greater than our own — a conscious, purposeful, rational, intelligent designer who's amazingly creative."*[12]

## THE EVIDENCE OF CONSCIOUSNESS

Human consciousness, our ability to think and reason and be self-aware, is a complete enigma to evolutionists. How can chemicals and molecules be capable of thought? Evolutionist philosopher, Michael Ruse, FRSC, puzzles over this very problem:

> *"Why should a bunch of atoms have thinking ability? Why should I, even as I write now, be able to reflect on what I am doing? And why should you, even as you read now, be able to ponder my points? ... No one, certainly not the Darwinianist [evolutionist],*

*seems to have an answer to this. The point is that there is no scientific answer."[13]*

Ruse is entirely correct. Science simply **cannot** explain human consciousness. Dr. Michael Polanyi, arguably the foremost philosopher of science in the 20th century, wrote:

*"Mental processes cannot be explained by physics and chemistry. The laws of physics and chemistry do not ascribe consciousness to any process controlled by them: the presence of consciousness proves, therefore, that other principles than those of inanimate matter participate in the conscious operations of living things."[14]*

Evolutionist scientists, in an attempt to explain how consciousness can arise from the human brain, have recently begun using several nebulous phrases:

- **Non-local consciousness:** The idea that our brains are merely the conduit for the mind, not the source of its origin.[15] Scientists who are exploring this concept are investigating what they refer to as "fringe phenomena", such as near-death experiences and precognition, in the hope of gaining an insight into consciousness.

- **Integrated Information Theory:** The idea that every molecule of biological life has a tiny amount of consciousness which, when combined together produces a higher level of combined consciousness. Some proponents of this theory even suggest that *all* matter, even *non-living* matter, has a tiny spark of consciousness.

- **Emergent Property theory:** This follows on from the previous idea. It proposes that the latent consciousness within all matter is activated by the particular complex arrangements of the human brain, which is able to activate and focus this consciousness more effectively than other animals or organisms.

What is striking about all of these theories is their frank admission that consciousness cannot be explained by mere physical chemicals and atoms. Each of these theories rests on the premise that consciousness has a non-physical source. Where these theories all fall down, however, is their inability to explain how a purely physical, mechanistic universe could have created a universal consciousness that somehow pervades each molecule of matter. Where did this consciousness come from? Evolutionists have absolutely no reasonable answer.

Human consciousness indicates that we are more than mere physical machines. We are more than flesh and blood. There is an esoteric essence within each person that transcends the merely physical. Our consciousness, our ability to think, and feel, and reason, is extremely convincing evidence for the existence of an intelligent designer who has implanted a soul within each of us.

Theologian and philosopher, Dr. J. P. Moreland, states:

> *"You can't get something from nothing. If the universe began with dead matter having no consciousness, how, then, do you get something totally different - conscious, living, thinking, feeling, believing creatures - from materials that don't have that? But if everything started with the mind of God, we don't have a problem with explaining the origin of our mind."*[16]

## CONCLUSION

Evolutionists such as Richard Dawkins will, no doubt, continue to claim that *"there is not a tiny shred of evidence for the existence of any kind of god."*[17] But for those with an open mind, the evidence for the existence of an intelligent designer is all around:

- The extraordinarily fine-tuned universal constants, which hold our universe together.

- Our miraculously shaped and strategically located planet that is perfectly designed for life.

- The unbelievably complex genetic information encoded into each of our cells.

- The irreducible complexity of the single cell.

- The origin of biological life.

- The mystery of human consciousness.

It is evidence such as this that led physicist, Dr Paul Davies, to conclude:

> "It may seem bizarre, but in my opinion, <u>science offers a surer path to God than religion</u>. People take it for granted that the physical world is both ordered and intelligible. The underlying order in nature - the laws of physics - are simply accepted as given, as brute facts. Nobody asks where they came from; at least they do not do so in polite company. However, even the most atheistic scientist accepts as an act of faith that the universe is not absurd, that there is a rational basis to physical existence manifested as law-like order in nature that is at least partly comprehensible to us. So science can proceed only if the scientist adopts an essentially theological worldview."

Having said this, you do not need to be a scientist to discern the evidence of God's existence. It is clear and evident to anyone who has an open mind. The beauty and complexity of the universe is a symphony that declares the greatness of the God who created it. Thus, the Apostle Paul, in his letter to the Romans, writes:

# Evidence For A Creator: Intelligent Design

*"What may be known about God is plain to them, because God has made it plain to them. For since the creation of the world God's invisible qualities—his eternal power and divine nature—have been clearly seen, being understood from what has been made, so that people are without excuse." (Romans 1:19-20).*

---

**ENDNOTES - Chapter 14**

---

[1] Walter Bradley, "The Just So Universe", in William A. Dembski and James M. Kushiner, "Signs of Intelligence", Baker Press, Grand Rapids, 2001, p.170

[2] Robin Collins, in an interview with Lee Strobel, transcript in Lee Strobel, "The Case for a Creator", Zondervan, Grand Rapids, 2004, pp.161-162

[3] William Lane Craig, "On Guard", David C. Cook, Ontario Canada, 2010, p.109

[4] William Lane Craig, "On Guard", David C. Cook, Ontario Canada, 2010, p.109

[5] Robin Collins, "The Teleological Argument", on http://citeseerx.ist.psu.edu/

[6] Paul Davies, "The Mind of God", New York, Touchstone, 1992, p.16

[7] https://en.wikipedia.org/wiki/Proxima_Centauri_b

[8] http://blogs.discovermagazine.com/d-brief/2016/02/22/earth-is-a-1-in-700-quintillion-kind-of-place/#.XFZp56ozZPY

[9] https://blogs.scientificamerican.com/cross-check/can-faith-and-science-coexist

[10] Michael Behe, quoted in Lee Strobel, "The Case for a Creator", Zondervan, Grand Rapids, 2004, p. 153ff

[11] Statement by Isaac Newton reported in Charles Dickens's "All the Year Round" (1864), Vol. 10, p. 346; later found in "The Book of the Hand" (1867) by A R. Craig, S. Low and Marston, p. 51:

[12] Lee Strobel, "The Case for a Creator", Zondervan, Grand Rapids, 2004, p.275f

[13] Miachel Ruse, in Lee Strobel, "The Case for a Creator", Zondervan, Grand Rapids, 2004, p.307

[14] Michael Polanyi, "The Anatomy of Knowledge", Rutledge Press, London, 1966 p.323

[15] https://www.gaia.com/article/consciousness-is-a-big-problem-for-science

[16] J. P. Moreland, in Lee Strobel, "The Case for a Creator", Zondervan, Grand Rapids, 2004, p.307 f.

[17] https://www.youtube.com/watch?v=of-8Q3HySjE&t=44m08s

# Evidence For A Creator: Historical and Personal

Chapter 15

# Evidence For A Creator: Historical Evidence and Personal Experience

This is a book primarily about scientific evidence; scientific evidence that contradicts evolution and scientific evidence that points to the existence of a Creator. But scientific evidence is not the only evidence for God's existence. In this chapter I want to briefly outline two other extremely important categories of evidence for the existence of God: historical evidence and personal experience. If you are simply seeking greater insight into the scientific arguments against evolution you may decide to skip this chapter. But for those who are wondering whether the possibility of God's existence is a viable alternative to the theory of evolution, this chapter will outline further extremely convincing evidence.

## HISTORICAL EVIDENCE

If God exists, surely we should expect to be able to discern his presence throughout the course of human history. Surely a God who cared enough to create such a magnificent universe, and who formed mankind so wonderfully, would not walk away and have nothing further to do with us.

Christians believe that God *has* intervened in human history, very clearly, at various times. Many of these interventions are carefully recorded for us in the Bible, the clearest of which is the appearance of Jesus Christ on our planet. You may be one of those people who is inclined to dismiss the stories of Jesus as myth, and regard the Bible as a fairy story, but I ask you to suspend your disbelief momentarily and objectively consider the evidence I am about to present.

The life, death and resurrection of Jesus Christ constitutes the strongest possible evidence for the existence of God. The four written Gospels, Matthew Mark, Luke and John, provide a detailed historical account of Jesus' miraculous life and, ultimately, his resurrection from the dead. Forty separate extraordinary miracles are recorded. These include healing the blind, healing quadriplegics and paraplegics, walking on water, instantly calming storms, multiplying food to feed thousands and raising three dead people back to life, one of whom had been dead for four days! Jesus claimed to be none other than God in the flesh, and he ultimately proved this by rising from the dead after having been dead and entombed for three days. If the biblical accounts of the events of his life are historically accurate, then the question, *"Does God exist?"* has been answered in the most emphatic, unequivocal way.

Of course, this raises a crucial issue: Is the biblical account of Jesus' life accurate? Does it reflect actual history or is it merely a fanciful embellishment of an ancient mythological figure?

**The Claim That Jesus Never Existed**

I am occasionally amazed to meet someone who believes that Jesus never existed. (Indeed, the noted atheist, Richard Dawkins, argued this for a period of time, until he was corrected by historians). With all due respect, the claim that Jesus never existed is an absurd position to hold that completely overlooks the considerable weight of historical evidence. In my experience, the only people who hold this view are those who have no historical training and have not bothered to examine the evidence. The life of Christ is substantiated by a number of extra-biblical (outside of the Bible) writers from antiquity, including Cornelius Tacitus, Lucian, Flavius Josephus, Suetonius, Pliny the Younger, Thallus, Philegon and Mara Bar-Serapion. (We will examine relevant quotes from them in the next section of this chapter).

Respected historian and professed atheist, Neil Carter, writes:

> *"I can't believe I'm feeling the need to do this, but today I'd like to write a brief defence of the historicity of Jesus. When people in the sceptic community argue that Jesus never existed, they are dismissing a large body of work for which they have insufficient appreciation, most often due to the fact that they themselves have never formally studied the subject.... The earliest writings which attest to the existence of Jesus come from the apostle Paul, a leather worker by day and preacher by night ... sometime in the mid-50s AD... The oral tradition which later came to inform the writing of the gospels predates the ministry of Paul by many years... Paul didn't invent these stories..."*[1]

Dr. Bart Ehrman is another respected historian and a professed agnostic (uncertain whether God exists). He was recently interviewed by "The

Atheist Guy" on "Atheist Radio" (an internet radio station whose sole aim is to discredit Christianity)[2]. Here is a transcript of part of that interview:

> **Atheist Guy:** *Do you believe that Jesus actually existed?*
>
> **Dr. Ehrman:** *Yes. There is no serious historian who doubts the existence of Jesus. There are a lot of people who want to write sensational books claiming that Jesus didn't exist, but I don't know any <u>serious scholar</u> who doubts the existence of Jesus.*
>
> **Atheist Guy:** *But there are historians who disagree with you, aren't there?*
>
> **Dr. Ehrman:** *None that I've ever heard of. Not serious historians. I know thousands of scholars of the ancient world and I don't know any one of these scholars who disagree.*

Of course, it is one thing to concede that Jesus existed, it is quite another to accept the accounts of his miracles.

**The Claim That Jesus' Life Was Significantly Embellished**

Recognising the hopelessness of trying to disprove Jesus' existence, the major thrust of the modern atheistic attack upon Jesus has been the attempt to discredit the veracity of the biblical accounts of his life. Sceptics argue that Jesus' life was significantly embellished by the New Testament writers in order to promulgate a new religion – a religion which his followers had invented. According to this theory, Jesus was just an ordinary man who said a few wise things and developed a popular following. The theory proposes that after his death, his followers deified him (declared him to be God), concocting stories of supposed miracles and fabricating the myth of his resurrection. The problem with this view is that it is overwhelmingly refuted by the Criteria of Historical Reliability.

**The Criteria of Historical Reliability**

In assessing whether an ancient document is historically reliable, historians apply multiple tests. These tests are commonly referred to as Criteria of Historical Reliability. The most commonly employed criteria are as follows:

1. The Criterion of Time Gap
2. The Criterion of Embarrassment
3. The Criterion of Multiple Attestation
4. The Criterion of The Absence of Protestation
5. The Criterion of Corroboration by Hostile Critics
6. The Criterion of Author Credibility

Space does not permit a full explanation of each of these, but a brief overview of several of the criteria will help you to understand why the New Testament letters, and the Gospels in particular, are widely regarded as historical documents of the highest calibre.

**The Criteria of Time Gap and The Absence of Protestation (Criteria 1 and 4)**

In historical terms, "time gap" refers to the amount of elapsed time between the actual historical events and the time of writing (when those events were eventually written down). In order for successful embellishment to occur, a significant amount of time has to elapse, so that there are very few, if any, eye-witnesses left alive who could refute the embellishment.

Professor A. N. Sherwin-White (1911-1993) was a world-renowned Greco-Roman historian who, among his many scholarly works, analysed ancient historical embellishments. He concluded that minor embellishment required a time gap of at least two generations, while major embellishment required at least 200 years.[3] In other words, even minor mythological embellishment can't gain traction if it is written within the lifespan of the first two generations after the event, because there are too many people still alive who know the facts and who could speak up to refute the embellishment.

Let me give you an example. At the time of this book's publication, it is 73 years since Winston Churchill led the Allies to victory over Adolf Hitler and the Nazis. Suppose I decided to write a biography of Churchill and I concocted all kinds of fanciful stories of him working miracles, raising people from the dead, walking on water and rising from the dead himself. I would very quickly be shut down as a fool and my written account would be overwhelmed by a flood of literary rebuttal. And in 2,000 years time, anyone investigating my ridiculous claims would uncover a veritable sea of indignant refutation. Even after more than 70 years have elapsed, I still could not get away with embellishing the life of Winston Churchill!

In the case of the life of Jesus Christ, the first three Gospels (Matthew, Mark and Luke) were written only 30-35 years after Jesus' death, and the Gospel of John was written another 25 years later. And these Gospels drew upon even earlier written accounts of the life of Jesus (that are no longer extant), some of which historians believe were written within the first few years of Jesus' death and resurrection. In other words, the Gospel stories were written within the lifetime of the eye-witnesses to the events of Jesus' life. And the response of the ancient world to those Gospels was remarkable; the Gospels were met with RESOUNDING SILENCE. No refutation. No protest. The reason for this is quite simple. No one could refute the Gospels because there were still thousands of eyewitnesses alive who had witnessed Jesus' miracles. Hundreds had witnessed him raise people from the dead. Thousands had witnessed him heal the sick, the blind, the paralysed. Thousands had witnessed him multiply food in order to feed the hungry. Even Jesus' death and resurrection from the dead was witnessed by hundreds. The Apostle

Paul, in 1 Corinthians 15:6, tells us that on one occasion Jesus appeared alive again to a crowd of more than 500 people. As the Gospels were written and distributed in first century Palestine, there were literally thousands of eyewitnesses who could verify their veracity. That is why we do not find any literary refutation in the historical record. If the Gospel accounts were fanciful embellishments by some deluded followers, we would expect to find a flood of literary refutation. But there is none.

Significantly, if the resurrection of Jesus was a myth, foisted upon the world by his over-zealous disciples, the Jewish authorities could have easily quashed the rumours by producing the dead body of Jesus. The fact that they did not do so is significant. In fact, the absence of protestation by the authorities is astounding. Not only did they fail to produce a dead body, but they also failed to write a single word refuting either the resurrection of Jesus or his miracles. Because they couldn't! The facts were simply irrefutable; Jerusalem and the whole of Judea and Galilee were full of eyewitnesses who could verify the events.

Both the extremely short time gap and the complete absence of protestation are two vital criteria in establishing the historical veracity of the Gospel accounts of the life of Jesus.

### The Criterion of Multiple Attestation (Criterion 3)

This test simply means that the more independent sources that report the same event, the more likely that event is to be historically accurate. In the case of the Biblical account of the life of Jesus, there is both internal and external multiple attestation. In terms of internal multiple attestation, many of the events of Jesus' life were recorded by multiple writers within the Bible itself: Matthew, Mark, Luke, John, James, Peter, Paul, John the elder and Jude. This is not to be undervalued, for these writers wrote from geographic and temporal isolation and their unified voice is particularly impressive.

# Evidence For A Creator: Historical and Personal

In terms of external multiple attestation, a number of events in the life of Jesus are corroborated by external sources:

- **Cornelius Tacitus**, a Roman senator, described the death of Jesus at the hands of Pilate.[4]

- **Gaius Suetonius Tranquillus**, a Roman Historian, referred to Christ as the leader of the sect of Christians that was plaguing Rome.[5]

- **Thallus**, a first century Greek / Roman historian, described the supernatural darkness and the earthquakes that accompanied the crucifixion of Christ (recorded in all three synoptic Gospels – Matthew, Mark and Luke). His reference is quoted in the early second century by the historian, Africanus; *"On the whole world there pressed a fearful darkness, and the rocks were rent by an earthquake, and many places in Judea and other districts were thrown down. Thallus calls this darkness an eclipse of the sun in the third book of histories, without reason it seems to me."*[6]

- **Tertullian**, a second century Christian writer, in his *"Apologeticus"*, told the story of the crucifixion darkness and suggested that the evidence must still be held in the Roman archives.[7]

- **Phlegon of Tralles**, a second century Greek historian, also recorded the supernatural darkness, accompanied by earthquakes felt in other parts of the Empire during the reign of Tiberius (probably that of 29 AD). He wrote, *"there was the greatest eclipse of the sun and it became night in the sixth hour of the day so that stars even appeared in the heavens. There was a great earthquake in Bithynia, and many things were overturned in Nicaea."*[8] He also wrote that *"Jesus, while alive, was of no assistance to himself, but that he arose after death,* *and exhibited the marks of his punishment, and showed how his hands had been pierced by nails."*[9]

**The Criterion of Corroboration by Hostile Critics (Criterion 5)**

The primary idea of this test is that hostile critics, (people who are outside of a particular belief system and who disagree with it), are the least likely to affirm or validate something that they are opposed to. Corroboration by hostile critics is, therefore, extremely persuasive in establishing the historical validity of a document or an event. In the case of the Biblical account of the life, miracles, death and resurrection of Jesus Christ, there are several important sources of hostile corroboration.

**Firstly**, there are the Roman and Greek historians already mentioned, none of whom (except for Tertullian) were Christians. As Romans and Greeks, they would have been philosophically opposed to the teachings of Christianity, yet they confirmed key events in the life of Jesus.

**Secondly**, and even more significantly, the Jewish sacred text, the Talmud, mentions Jesus on several occasions, even referring to his miracles. It attributes these miracles to the power of the devil and accuses Jesus of *"sorcery"*. For example, The Babylonian Talmud (BT, Sanhedrin, 43a) states:

*"On the eve of the Passover Yeshua [the Nazarene] was hanged. For forty days before the execution took place a herald went forth and cried, "He is going forth to be stoned because he has practised sorcery and enticed Israel to apostasy. Anyone who can say anything in his favour, let him come and plead on his behalf." And since nothing was brought forward in his favour, he was hanged on the eve of Passover."*[10]

Significantly, even Jesus' enemies, the Jews, could not deny the power of his miracles, instead choosing to attribute them to evil sorcery.[11]

**Thirdly**, there is the famous reference by Flavius Josephus, a practising Jew and an outstanding first century historian. In 93 AD, in Rome, Josephus published his lauded work, *"Antiquities of the Jews"*, which included the following account;

> *"About this time there lived Jesus, a wise man, if indeed one ought to call him a man. For he was one who performed surprising deeds and was a teacher of such people as accept the truth gladly. He won over many Jews and many of the Greeks. He was the Messiah. And when, upon the accusation of the principal men among us, Pilate had condemned him to a cross, those who had first come to love him did not cease. He appeared to them spending a third day restored to life, for the prophets of God had foretold these things and a thousand other marvels about him. And the tribe of the Christians, so called after him, has still to this day not disappeared."*[12]

This passage has been embroiled in controversy since the 1700's, with atheists and sceptics arguing that the passage must be an interpolation by later Christian copyists, as a Jew would surely never have written such a favourable report of Christ. The ensuing debate throughout the subsequent centuries has generated the equivalent of a small library of academic papers and books, either defending or refuting the authenticity of this one small paragraph. In 1995, however, a discovery was published that brought important new evidence to the debate. It uncovered an earlier document that Josephus had used as a source document for his comments about Jesus, thus proving that they were not added by a later copyist.[13] While aggressive atheists still maintain that this paragraph must have been inserted by a Christian copyist, the vast majority of neutral academics are now of the opinion that the reference is largely authentic (with the possible exception of the words *"if indeed one ought to call him a man"* and *"He was the Messiah"*, which are still disputed). This reference by Josephus represents a powerful corroboration by a hostile critic of the miracles of Jesus and the historicity of his resurrection.

**The Resurrection of Jesus: The Ultimate Proof**

The greatest proof of Jesus' claim to be God was, of course, his resurrection from the dead. If his claim of divinity had had no substance, if it had arisen from mental delusion or deliberate fraud, the grave would have been the end of the matter, putting an end to his claims conclusively. But the Bible records Jesus' resurrection from the dead as the ultimate proof of his divinity. According to the Gospel accounts, Jesus rose to life and appeared to his followers over a period of 40 days, with Paul mentioning one occasion when he appeared to a crowd of over 500 people (1 Cor 15:1-11). It also records the inability of the Jewish authorities to produce his dead body in order to quash the news of his resurrection, as well as their failed attempt to spread the rumour that Jesus' body had been stolen by his disciples (Matt 28:13).

Furthermore, there is the compelling evidence of the unwillingness of the Apostles and other disciples to recant their belief in the resurrection of Jesus, even when it meant facing death for that belief. Of the twelve Apostles appointed by Jesus, eleven were ultimately put to death for refusing to cease proclaiming their message about the resurrection and divinity of Jesus. Many hundreds of other first century Christians were also put to death at the hands of both the Jews and the Romans, because of their unwillingness to renounce this belief. Nobody is willing to die for something they _know_ to be a lie; at least no one who is sane! If the resurrection was not true, if it was a fabrication invented by the disciples, they would not have been willing to die for it. The fact that the Apostles and many other first century Christians went willingly to their deaths at the hands of the governing authorities, refusing to renounce their belief in Christ's resurrection, is extremely compelling evidence for its authenticity.

# Evidence For A Creator: Historical and Personal

*"OK guys - here's the plan! We get the body out of the tomb and stash it somewhere. Then we come back here and tell a story that will probably get us all killed. So who's with me on this?"*

**Sceptics Convinced and Converted by The Resurrection**

If the accounts of the resurrection of Jesus are true, then we have incontrovertible evidence of the existence of God. Realising this, several academics and historians have, over the years, set out to disprove the resurrection story. Their philosophy was simple; disprove the resurrection and you remove the strongest argument for the existence of God. Not only have these attempts been unsuccessful, they have regularly led to the complete capitulation and conversion of those undertaking the research:

- **Sir William Mitchell Ramsay** (1851-1939) was a highly respected historian and archaeologist from Scotland. He set out to prove the historical inaccuracies of Luke and Acts. He spent 15 years researching and digging, only to end up being convinced of the incredible accuracy of the New Testament. He converted to Christianity, and called Luke one of the greatest historians to ever live. He wrote several books on the subject, which have stood the test of time. His work caused an outcry from atheists because they had been funding his research and were eagerly awaiting his results in disproving the validity of the New Testament![14]

- **Albert Henry Ross** (1881-1950) was an English journalist and author who set out to disprove the "myth" of the resurrection. He was planning on writing a paper called *"Jesus – The Last Phase"*, but he became converted during the course of his investigations. He ended up writing the classic book *"Who Moved The Stone?"*[15] under the pseudonym Frank Morrison. The book has led many people to faith in Christ.

- **Lee Strobel** (born 1952) was a journalist for the Chicago Tribune. His wife converted to Christianity and Lee became very concerned. In order to "rescue" his wife from the church, he set out to disprove Christianity, focussing on the resurrection story. He spent 18 months, utilising his skills as a researcher and investigative reporter, interviewing experts from around the world, and studying the 1st century documents for himself. The overwhelming evidence for the resurrection of Christ eventually led to his own conversion, and he went on to write the now famous book, *"The Case For Christ"*[16] (recently made into a movie), along with several other books in the series, which have led many people to faith in Christ.

- **Simon Greenleaf** (1783 - 1853) was a professor of law at Harvard University and a towering figure in the legal world. He set out to disprove the Gospel accounts, particularly the narratives concerning the resurrection of Jesus. He rigorously applied the criteria of historical reliability to the Gospels to test their authenticity, and applied cross-examination legal techniques to assess the reliability of the eye-witness accounts. Eventually he, too, was converted to Christianity, and his rigorous approach was foundational in the development of modern juridical Christian apologetics.[17]

- **C. S. Lewis**, (1898 - 1963) was a distinguished professor of English literature at both Oxford and Cambridge universities (at different times), and is widely regarded as one of the greatest minds of the 20th century. He was originally an ardent atheist who fought long and hard to maintain his disbelief. In his book, "Surprised by Joy", he wrote of his eventual conversion, stating, *"In the Trinity Term of 1929 I gave in, and admitted that God was God, and knelt and prayed: perhaps, that night, the most dejected and reluctant convert in all England."*[18] Lewis's inability to refute the solid historical evidence for the life and resurrection of Jesus led to his reluctant conversion – a conversion that would eventually result in the publication of, arguably, the most profound Christian philosophical writings of the modern era.

It is the reluctance of these learned scholars to be converted, that makes their eventual conversions all the more significant. Indeed, history is replete with respected scholars who set out to disprove the resurrection of Jesus and were eventually convinced by the overwhelming weight of historical evidence.

The miraculous life and ultimate resurrection of Jesus Christ is the strongest possible evidence for the existence of God, recording his extraordinary appearance on our planet over 2,000 years ago. While we may wish God to be as obvious to us today as he was in the life of Jesus, we simply cannot ignore the fact that, at one point in history, God revealed himself to humanity in the clearest possible manner. The life of Jesus is extremely compelling evidence for the existence of God.

**PERSONAL EXPERIENCE**

The personal experience of millions of Christians today must also be seriously considered. Many Christians testify that they have encountered God in a profound and personal way. This is in accord with God's promise in the Bible:

> *"You will seek me and find me when you seek me with all your heart. I will be found by you, declares the Lord." (Jeremiah 29:13-14)*

God promises to interact with us personally if we approach him with a humble, sincere heart. This has been my own experience. I experience God's presence in my life on a daily basis in subtle ways, and, occasionally, in very profound ways.

What is to be made of this kind of experiential claim by millions of Christians? Obviously, just because people claim to experience something does not necessarily mean that their experience reflects reality. People can easily be self-deluded. Even the fact that large numbers of people attest to experiencing God's presence does not prove the validity of their experience. Large numbers of people can be wrong. There are people who are convinced that they have experienced all kinds of things: abduction by aliens, communication with aliens, the magical power of crystals, communication with the dead, the ability to leave their bodies and astro-travel.

On the other hand, personal testimony is considered to be a powerful legal proof, when it is substantiated by other verifiable evidence. In a court of law, the testimony of just two corroborating eyewitnesses is often sufficient to obtain a clear verdict.

The testimony of Christians regarding their encounter with God must be considered very seriously on a number of grounds:

- **Volume:** Not just hundreds, nor thousands, but millions of Christians today and throughout the ages testify to experiencing God in a profound way in their lives.

- **Corroboration:** There is overwhelming corroboration in the testimony of Christians regarding their experience of the presence of God in their lives. Their experience of God's protection, guidance, empowering, peace and joy, together with countless testimonies of answered prayer and miraculous interventions, is a particularly impressive homogenous body of evidence.

- **The Character and Integrity of the Witnesses**: In a court of law, the character of the eyewitnesses will determine the weight that is given to their testimony. Unstable "crack pots" are not usually even given a hearing, whereas the testimony of sane, intelligent, coherent eyewitnesses is accorded great weight. Among the Christian community who claim to experience the presence of God in their lives, there are highly intelligent, sane, rational, clear thinking people, not given to wild imaginative delusions. They are drawn from all walks of life, from straight-shooting manual labourers to highly qualified scientists, doctors, lawyers and other professionals.

While the testimony of millions of Christians who claim to experience God personally may not be acceptable to the sceptic as definitive proof of God's existence, it cannot not be lightly dismissed. The challenge to sceptics, is whether they are willing to open their own hearts to experience the presence of God in their lives. Because God has invited us all to do so:

> "Taste and see that the Lord is good." (Psalm 34:8)

And God has given all mankind a clear promise:

> "You will seek me and find me when you seek me with all your heart. I will be found by you, declares the Lord." (Jeremiah 29:13-14)

# Evidence For A Creator: Historical and Personal

## ENDNOTES - Chapter 15

[1] Neil Carter, quoted inpatheos.com/blogs/godlessindixie/2014

[2] https://www.youtube.com/watch?v=u9CC7qNZkOE

[3] "Myth Growth Rates and The Gospels", http://www.bibleinterp.com/articles/2013/kom378030.shtml

[4] Cornelius Tacitus, "Annals" (written ca. AD 116), book 15, chapter 44

[5] Suetonius, "Lives of the Twelve Caesars", 121 AD

[6] Africanus, quoting Thallus (96 AD), https://en.wikipedia.org/wiki/Thallus_(historian)

[7] Tertullian, "Apologeticus", (197 AD), quoted in https://en.wikipedia.org/wiki/Crucifixion_darkness

[8] Plegon of Tralles, quoted by Origen of Alexandria (182-254 AD), in Against Celsus (Book II, Chap. XIV),

[9] Plegon of Tralles, quoted by Origen of Alexandria (182-254 AD), in Against Celsus (Book II, Chap. XIV),

[10] BT, Sanhedrin 43a, quoted in https://en.wikipedia.org/wiki/Yeshu#Yeshu_the_sorcerer

[11] BT, Sanhedrin 43a, quoted in https://en.wikipedia.org/wiki/Yeshu#Yeshu_the_sorcerer

[12] Flavius Josephus, "Antiquities of The Jews", 18.3.3 §63

[13] "The Testimonium Flavianum", http://www.josephus.org/testimonium.htm

[14] Josh McDowell, "Evidence That Demands A Verdict", Thomas Nelson Inc, 1979

[15] Frank Morrison, "Who Moved The Stone", Zondervan Press. 1930 (latest edition 2002)

[16] Lee Strobel, "The Case For Christ", Zondervan Press, 1998

[17] https://en.wikipedia.org/wiki/Simon_Greenleaf

[18] C. S. Lewis, "Surprised By Joy", Harper Collins, New York, 1955, Ch. 14, p. 266

# The Light From The Stars

Chapter 16

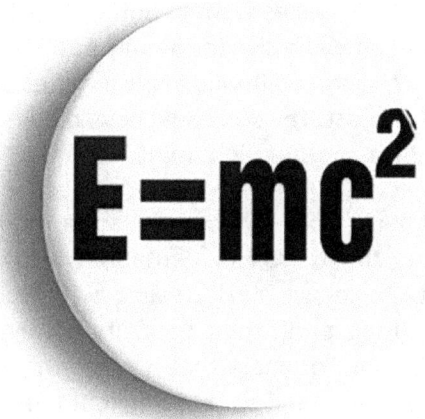

# The Light From The Stars

Prepare to have your mind blown! We are about to examine an issue that is truly mind-boggling, but I will attempt to keep it as simple as possible.

**THE STAR LIGHT PROBLEM**

For those (like me) who believe in a literal six-day creation, as described in the first two chapters of Genesis, the light from the stars poses a theoretical problem. If the Genesis account is correct and the genealogies of the Bible are accurate, then the whole universe was created only about 6,000 years ago. But the light from the stars appears to tell a different story.

# The Light From The Stars

Let me explain. The speed of light is 299,792.458 km per second. Let's round it up to 300,000 km per second. At that speed, a beam of light can travel approximately eight times around the earth in a single second. Our sun is 149.6 million km from the earth. The light from the sun, travelling at 300,000 km per second, takes eight minutes to reach Earth. Therefore, when we look at the sun in the sky, (hopefully not directly with the naked eye), we are seeing the sun as it was eight minutes ago. We are seeing eight minutes into the past. The sun could have blown up one second ago and we won't see it for another eight minutes!

But the sun is relatively close by cosmic standards. The closest star to earth, apart from our sun, is Proxima Centauri, which is 40 million million kilometres distant. Once we start dealing with such vast distances, cosmologists start using a different measure of distance, called light-years. This is the amount of years that it takes the light from a star to reach Earth. Proxima Centauri is so far away that the light that it radiates, travelling at 300,000 km per second, takes 4.3 years to reach Earth. Thus, cosmologists say that it is 4.3 light-years from Earth. This means that when we look at Proxima Centauri through a telescope, the light reaching our eyes left that star 4.3 years ago. We are seeing the star as it was 4.3 years ago, and we are looking 4.3 years into the past.

But Proxima Centauri is an extremely close star. The universe is unimaginably vast. The furthest stars and galaxies that we have been able to view through our most powerful telescopes are about 14 *billion* light years away. Theoretically, this seems to indicate that the light from those galaxies has taken 14 billion years to reach Earth. We are apparently looking back in time 14 billion years! But how can this be, if the Bible indicates that the universe is only thousands of years old?

Not only does the starlight issue pose a potential problem for six-day creationists, but, surprisingly, it also creates a problem for big bang evolutionists. Measurements of ambient temperatures of the cosmos (through spectral analysis of various emissions) show a very even level of cosmic microwave background radiation (CMBR) spread throughout the entire universe. In other words, our observation of the ambient temperature of the outer reaches of the universe (supposedly billions of years ago) are close to the same temperature as the ambient temperature of nearby space (in the present).[1] But this should not be the case if there was a "big bang". The universe would have had a much higher ambient temperature immediately after the big bang and would have gradually cooled with the passage of time, resulting in present day observations of the outer edges of the universe (with light from billions of years ago) showing much higher temperatures than the temperature of present day, nearby space.

The even spread of cosmic ambient temperature is a huge problem for big-bang evolutionists who maintain that when we look at distant galaxies, we are looking billions of years into the past. This is referred to as the "horizon problem". An article in the March 2005 issue of New Scientist, stated:

> *"This horizon problem is a big headache for [evolutionist] cosmologists."*[2] *[Brackets mine].*

## A STAR LIGHT SOLUTION

The simplistic explanation, offered by some biblical creationists, is that when God created the stars on Day 4 of creation, he also created the light in transit, already reaching to the earth. Thus, Adam could immediately see the stars without having to wait millions and billions of years for each star to pop into visual existence. The problem with this idea is that, if this was the case, the light we receive now from distant galaxies would portray movement and events that never actually happened; it would be a false light show, effectively deceiving us. For example, if we look into the night sky tonight and see a star that is 100 million light years distant going supernova (exploding), that event would not have really happened, because it is just part of the original light beam that God instantaneously created reaching the earth. It would take 100 million years before we see what is really happening with that star. Since God is not a God of deception, but a God of truth (Num 23:19 and Heb 6:18), this does not sit well with many people.

A number of recent studies, however, have opened up the possibility of a more reasonable and consistent explanation. Newly emerged evidence is radically changing the way scientists view the relationship between gravity, velocity, speed of light and the passage of time. There are two scientific developments and one biblical concept that need to be combined in order to reach a workable hypothesis for the starlight problem.

**Gravitational Time Dilation**

Einstein's general theory of relativity indicates that the faster an object travels, the more slowly time passes for that object relative to the rest of the universe. Thus time appears to slow down (from an outsider's perspective) as velocity increases. Similarly, Einstein's theory also predicts that time slows down when an object is subject to increased gravity or continual gravitational acceleration. This is referred to as "gravitational time dilation".

This speeding up of time is not just theoretical; it has been observed in real life. The earth's GPS navigational satellites, orbiting at an altitude of approximately 20,000 km, are subject to slightly less gravitational acceleration than we are on the surface of the earth. The atomic clocks on these satellites consistently run 38 microseconds per day **faster** than the same clocks on earth![3] This is due to the relativistic effect of gravitational time dilation. It might not seem that 38 microseconds per day is a big deal, but even that small discrepancy, if allowed to accumulate, would render our GPS system completely useless, because positional errors would build up at a rate of about 400 metres per hour.[4] Because of this, highly precise coding continuously updates the atomic clocks on these satellites, adjusting for gravitational time dilation, to ensure that the clocks do not get out of sync with earth-based clocks.

# The Light From The Stars

Dr John Hartnett, in his book, *"Starlight, Time and the New Physics"*,[5] explains that gravitational time dilation would also be extremely severe when rapid acceleration of the fabric of space takes place in an expanding universe, as was the case when the heavens were "stretched out" at the beginning of time.

Now, hold that thought while we briefly examine another one.

## The Galactocentric Universe

A second piece of the puzzle is a recent cosmological observation. Until recently, cosmologists have maintained that the earth is nothing special, and it certainly isn't the centre of the universe. But there is growing evidence that our position in the universe may be very close to the centre. Recent astronomical surveys[6] have measured the redshift of around 250,000 observable galaxies, indicating that they are expanding outwards from our Milky Way Galaxy. Dr. Mark Harwood interprets this data:

> *"A straightforward interpretation of this data is that the galaxies are distributed with a spherical shell-like symmetry with the Milky Way galaxy at or near the centre!"*[7]

## God Stretched Out the Heavens

The third piece of the puzzle is a biblical concept. The first chapter of Genesis indicates that God created the sun, moon and stars on day 4 of the creation week. There are at least 11 references in the Bible to God *"stretching out the heavens"* when he created the stars (e.g. Job 9:8; Isa 40:22). Furthermore, Genesis 1:15 indicates that the creation of the stars and celestial bodies was completed **within** Day 4 of creation; *"And it was so"*. This stretching out of the heavens was completed in a single 24-hour period from Earth's perspective.

## Putting It All Together

# The Light From The Stars

Taking all three of these concepts into account - the speeding up of time due to gravitational time dilation, the fact that the universe appears to be receding away from us in every direction, and the biblical description that God "stretched out the heavens" when he created the universe - allows us to reach a reasonable explanation for the fact that light from stars that are billions of light years away is visible to us on earth only 6,000 years after creation. When God created the stars, he stretched out the very fabric of the space-time continuum, on Day 4 of creation (presumably to a position fairly close to their present distance from earth). This would have involved unimaginably rapid acceleration to reach their current positions within just 24 hours. Dr. Mark Harwood proposes[8] that the gravitational forces and other phenomena generated by this stretching out would have been enormous, and these forces, in turn, could have resulted in massive gravitational time dilation. The "clocks" of the galaxies being dispersed could have run unthinkably faster than the "clocks" on earth. While only 24 hours passed on earth, billions of years would have passed for those galaxies being stretched to the outer edges of the universe. Dr. Mark Harwood explains:

> *"By the end of Day 4, when God completed his work of creating the sun, moon and stars, and had stretched out the heavens to their vast extent, billions of years of cosmic time could have elapsed at the outer edges of the cosmos in just one 24-hour earth day. There would have been more than enough time for the light from distant stars to have reached the earth so that when Adam gazed at the night sky on that sixth night, he would have seen much the same as what we see today."*[9]

So, what can we surmise when we view the light from stars and galaxies that are billions of light years away? **Firstly,** remember that the term "light-years" is a measure of distance, not of actual time. **Secondly,** the gravitational time dilation effect, in accordance with Einstein's general theory of relativity and confirmed by measurable observations, can easily account for the apparent age of the distant universe.

# The Light From The Stars

Thus, Dr Harwood comments:

> *"6,000 years have passed since the Creation Week. If the models outlined above are correct, the light we see today from any star that is greater than 6,000 light years away from the earth will have originated on Day 4 itself. This would include most of the visible stars, all of which are part of the Milky Way galaxy. We are effectively looking at God's creative activity on Day 4 as we gaze into the universe!"*[10]

It must be stressed that the above is only one of four current working hypotheses for how light from distant stars can have reached a 6,000-year-old earth. This model is called the Gravitational Time Dilation model. The other three models are:[11]

1. **Mature Creation:** God simply created the light already reaching the earth.

2. **CDK:** The speed of light is not constant as Einstein's theory stipulates, but has been slowing down since the beginning of creation. Secular scientists have even recently entertained this idea for their own purposes.

3. **Anisotropic Synchronous Conventions:** Light could have travelled to earth at almost infinitely fast speeds in the initial stages of creation.

For further reading on this topic, you might like to try Dr. John Hartnett's book, *"Starlight, Time and the New Physics"*.[12] If you have a headache after all this, you might like to try 2 paracetamol and a glass of water!

**ENDNOTES - Chapter 16**

[1] Carl Wieland, "Starlight and Time; A Further Breakthrough", in "Creation" magazine, **30**(1):12–14, December 2007

[2] New Scientist magazine, 19 March 2005

[3] https://creation.com/how-can-distant-starlight-reach-us-in-just-6000-years

[4] https://creation.com/how-can-distant-starlight-reach-us-in-just-6000-years

[5] John Hartnett, "Starlight, Time and the New Physics", 2007, Creation Ministries International,

[6] The Two Degree Field Galactic Redshift Survey, and the Sloane Digital Sky Survey, cited in https://creation.com/how-can-distant-starlight-reach-us-in-just-6000-years

[7] Mark Harwood, "How Can Distant Starlight Reach Us In Just 6,000 Years?, at https://creation.com/how-can-distant-starlight-reach-us-in-just-6000-years

[8] Mark Harwood, "How Can Distant Starlight Reach Us In Just 6,000 Years?, at https://creation.com/how-can-distant-starlight-reach-us-in-just-6000-years

[9] Mark Harwood, "How Can Distant Starlight Reach Us In Just 6,000 Years?, at https://creation.com/how-can-distant-starlight-reach-us-in-just-6000-years

[10] Mark Haywood, "How Can Distant Starlight Reach Us In Just 6,000 Years?, at https://creation.com/how-can-distant-starlight-reach-us-in-just-6000-years

[11] Eric Hovind, "How Could Light Have Travelled Millions of Years?; https://creationtoday.org/how-could-light-travel-millions-of-years/

[12] John Hartnett, "Starlight, Time and the New Physics", Creation Ministries International, 2010.

## Chapter 17

# The Tide Is Turning

Although the popular impression conveyed by the media is that evolution is proven and the scientific community is in accord on the matter, this is far from the truth. There is a growing number of scientists who have abandoned the theory completely, with many more who are at least seriously questioning its validity. This is not because there has been a religious revival among the world's scientists. It is simply a result of the mounting scientific evidence directly contradicting the theory.

In 2001, Australian scientist, Dr. John Ashton, published the book, "*In Six Days: Why Fifty Scientists Choose To Believe in Creation*".[1] The book contains 50 chapters, each written by a different scientist, explaining why they have rejected the theory of evolution in favour of the creationist model. The scientists are all highly credentialed, Ph.D. scientists, drawn from a wide range of scientific disciplines, some of whom hold senior research positions at major universities.

# The Tide Is Turning

The scientists who contributed to Dr. Ashton's book in 2001 are just the tip of the iceberg. Despite Wikipedia's claim that *"Nearly all of the scientific community accepts evolution as the dominant scientific theory of biological diversity"*[2], the growing tide of scientists abandoning the theory is accelerating. In 2006, the website, "Evolution News", published an article reporting the following:

> *"Over 500 doctoral scientists have now signed a statement publicly expressing their skepticism about the contemporary theory of Darwinian evolution. The Scientific Dissent From Darwinism statement reads: 'We are sceptical of claims for the ability of random mutation and natural selection to account for the complexity of life. Careful examination of the evidence for Darwinian theory should be encouraged'."*[3]

Significantly, the signatories to the published list of dissenters, were not merely science teachers with bachelor's degrees (no offense intended to our wonderful science teachers!), but were prominent Ph.D. scientists from around the world. The article continued:

> *"The list of 514 signatories includes member scientists from the prestigious US and Russian National Academy of Sciences. Signers include 154 biologists, the largest single scientific discipline represented on the list, as well as 76 chemists and 63 physicists. Signers hold doctorates in biological sciences, physics, chemistry, mathematics, medicine, computer science, and related disciplines. Many are professors or researchers at major universities and research institutions such as MIT, The Smithsonian, Cambridge University, UCLA, UC Berkeley, Princeton, the University of Pennsylvania, the Ohio State University, the University of Georgia, and the University of Washington."*

The publication of this list of dissenters coincided with the launch of the website, DissentfromDarwin.org, which invites scientists from around the world to add their names to the list of dissenters. Significantly, the website invites only those scientists with a Ph.D. to join the list. Their invitation reads as follows:

> *"If you have a Ph.D. in engineering, mathematics, computer science, biology, chemistry, or one of the other natural sciences, and you agree with the following statement, 'We are sceptical of claims for the ability of random mutation and natural selection to account for the complexity of life. Careful examination of the evidence for Darwinian theory should be encouraged,' then please contact us"*[4]

By 2016, the list of dissenters had grown to 1,035 scientists! And the list is still growing.

The original impetus for the list and for the establishment of the website was the ridiculously exaggerated claim by America's PBS television network as it launched a documentary series, entitled *"Evolution"* in 2001, which claimed:

> *"virtually every scientist in the world believes the theory to be true."*[5]

One of the founders of the Dissenters website, Dr. John G. West, associate director of Discovery Institute's Center for Science & Culture, Seattle, stated:

> *"Darwinists continue to claim that no serious scientists doubt the theory and yet here are 500 scientists who are willing to make public their scepticism about the theory."* [6]

One of the original signers of the list, Dr. David Berlinski, mathematician and science philosopher with Discovery Institute's Centre for Science and Culture, has stated:

> *"Darwin's theory of evolution is the great white elephant of contemporary thought. It is almost completely useless, and the object of superstitious awe."[7]*

Let us not exaggerate the level of dissent, however. Scientists who have rejected the theory of evolution are not yet in the majority, but they are no longer a tiny minority. There are no reliable statistics, but given the huge amount of debate within scientific circles currently taking place, those who no longer subscribe to evolution appear to be a very significant and growing segment of the scientific community.

As scientific disbelief in evolution mounts, evolutionists seem to be becoming increasingly strident in their defence of their dwindling theory. Why is this?

That is the topic of the next chapter.

---

**ENDNOTES - Chapter 17**

---

[1] John Ashton, "In Six Days", Masterbooks, Green Forest, AR, 2001, 2nd edition: 2010

[2] https://en.wikipedia.org/wiki/Level_of_support_for_evolution

[3] "Over 500 Scientists Proclaim Their Doubts About darwin's Theory of Evolution", https://evolutionnews.org/2006/02/over_500_scientists_proclaim_t/

[4] https://evolutionnews.org/2006/02/over_500_scientists_proclaim_t/

[5] http://www.pbs.org/wgbh/evolution/

[6] John G. West, quoted in "Over 500 Scientists Proclaim Their Doubts About darwin's Theory of Evolution", https://evolutionnews.org/2006/02/over_500_scientists_proclaim_t/

[7] David Berlinski, quoted in, "Over 500 Scientists Proclaim Their Doubts About Darwin's Theory of Evolution",
https://evolutionnews.org/2006/02/over_500_scientists_proclaim_t/

# The Atheistic Agenda

# Chapter 18

# The Atheistic Agenda

Underlying the passionate belief in evolution by many scientists, and their increasingly strident defence of it as a theory, is a distinctly unscientific philosophy; atheism. Richard Dawkins openly admitted this when he wrote:

> "Darwin made it possible to be an intellectually fulfilled atheist."[1]

For those who are determined to reject the concept of a creator God, evolution offers the only reasonable alternative.

Thus, zoologist Dr. D.M. Walton once wrote:

> "Evolution is a theory universally accepted, not because it can be proved to be true, but because the only alternative, 'special creation', is clearly incredible."[2]

We have already discussed how such a claim, that belief in a creator God is *"clearly incredible"*, is a completely unscientific statement to make, as God lies beyond the parameters of science to detect and study. But the above sentiment demonstrates how tenaciously evolutionists will cling to their theory *"not because it can be proved to be true"* (as Dr Walton freely admits!), but simply because **they want to believe it**!

Prominent atheist, Thomas Nagel, writes:

> *"I speak from experience, being strongly subject to this fear myself<u>: I want atheism to be true and am made uneasy by the fact that some of the most intelligent and well-informed people I know are religious believers</u>. It isn't just that I don't believe in God and, naturally, hope that I'm right in my belief. <u>It's that I hope there is no God! I don't want there to be a God</u>; I don't want the universe to be like that. My guess is that this cosmic authority problem is not a rare condition and that it is responsible for much of the scientism and reductionism of our time. One of the tendencies it supports is <u>the ludicrous overuse of evolutionary biology to explain everything about human life, including everything about the human mind</u> .... This is a somewhat ridiculous situation .... <u>it is just as irrational to be influenced in one's beliefs by the hope that God does not exist</u> as by the hope that God does exist."*[3][Underlining mine]

Nagel's frank admission offers a candid glimpse into the driving force behind evolutionary theory and the reason for the strident dogmatism that often characterises its proponents. The intractable promotion of evolution is fundamentally driven by an overwhelming desire for there to be no God. Inherent in this, is the desire to not be accountable to an ultimate moral judge; the desire to be a free moral agent, able to follow one's desires without fear of ultimate consequence and retribution.

Philosopher, Dr Edward Feser, comments:

> "A desire to be free of traditional moral standards, and a fear of certain consequences of the truth of religious belief, can also lead us to want to believe that we are just clever animals with no purpose to our lives other than the purposes we choose to give them, and that there is no cosmic judge who will punish us for disobeying an objective moral law."[4]

But it gets worse! Many evolutionists are choosing to cling to the theory, even though they **know** it to be scientifically impossible. Consider this astonishing admission by Dr. George Wald:

> "There are only two possibilities as to how life arose. One is spontaneous generation arising to evolution; the other is a supernatural creative act of God. There is no third possibility. Spontaneous generation, that life arose from non-living matter was scientifically disproved 120 years ago by Louis Pasteur and others. That leaves us with the only possible conclusion that life arose as a supernatural creative act of God. <u>I will not accept that philosophically because I do not want to believe in God. Therefore, I choose to believe in that which I know is scientifically impossible; spontaneous generation arising to evolution.</u>"[5]
> [Underlining mine]

In other words, "*I know evolution doesn't make scientific sense, but I will believe it anyway, because I don't want to consider that God may exist.*" Extraordinary!

## THE UGLY MILITANCY OF EVOLUTION

If all that creationists had to contend with was evolutionist scientists clinging to a dying theory, one could almost sympathise with the atheist's dilemma; like a forlorn group of shipwreck survivors adrift on the ocean in a rubber dingy with a large hole! But many evolutionists display an ugly aggression towards those who believe in a Creator. For example, in 2009, Richard Dawkins posted an ugly comment on his website. He eventually removed it, but, unfortunately for Dawkins, the damage was already done. The quote was picked up by various critics, and is published in a number of online articles:

> "If they've [the creationists] been told that there's an incompatibility between religion and evolution, well, let's convince them of evolution, and we're there! Because after all, we've got the evidence. ... I suspect that most of our regular readers here would agree that ridicule, of a humorous nature, is likely to be more effective than the sort of snuggling-up and head-patting that Jerry [Coyne] is attacking. I lately started to think that we need to go further: <u>go beyond humorous ridicule, sharpen our barbs to a point where they really hurt</u>. ...You might say that two can play at that game. Suppose the religious start treating us with naked contempt, how would we like it? I think the answer is that there is a real asymmetry here. <u>We have so much more to be contemptuous about! And we are so much better at it. We have scathingly witty spokesmen of the calibre of Christopher Hitchens and Sam Harris. Who have the faith-heads got, by comparison? Ann Coulter is about as good as it gets. We can't lose!</u>"[6] [Underlining mine]

In televised interviews Dr. Dawkins presents himself as a calm, reasonable, polite person, but in comments like this, he reveals the ugly hatred of religious belief that lies just beneath surface.

In recent years we have witnessed an increasingly militant campaign by atheistic evolutionists to not only defend their theory, but to aggressively promote it within the media and to persecute those within the scientific community who voice any kind of dissention. The influence of this atheistic movement extends to many universities and many forms of media. A recent example helps to illustrate the point.

Dr. Günter Bechly, a respected German palaeontologist, had a successful academic career. He had published numerous papers in prestigious, peer-reviewed journals, and was a curator at Stuttgart's State Museum of Natural History. All was going well for him until 2016, when he publicly declared his rejection of the theory of evolution. He became affiliated with the Discovery Institute, an online hub of intelligent design advocates, aimed at defending the concept of intelligent design and pointing out the flaws in evolutionary theory. He also took part in the making of a documentary film entitled "Revolutionary" that presented testimonies of scientists who are abandoning the theory of evolution.[7]

Evolutionists took exception to Dr. Bechly's public dissent regarding evolution, and placed pressure on the editors of Wikipedia to delete his Wikipedia page, on the grounds that he could no longer be considered a serious scientist. In October 2016, Wikipedia complied, and removed his page.[8] The extraordinary thing about this episode is that Wikipedia has decided that Dr. Bechly can no longer be considered a "proper" scientist, solely because he now publicly admits belief in creation. This is despite the fact that **he has several species named after him** and also that he has a **high academic citation ranking** (in what is called the h-index).[9] (Dr. Bechly now has a page on the German Wikipedia site, de.wikipedia, but still, at the time of this book's publication, no page on the international English Wikipedia site).

Sadly, Dr. Bechly's experience is not unique. In the opening chapter of this book, I cited the recent case of Dr. Gavriel Avital, the chief scientist for the Israeli Education Ministry, who was sacked for voicing his doubts concerning the validity of the theory of evolution.[10] Similarly, the 2008 documentary film, *"Expelled: No Intelligence Allowed"*[11], showed interviews with a number of scientists who had been sacked from various positions because they expressed their disbelief in evolution.

Not only is the academic world rife with prejudice against those who dare to speak up against evolution, but the atheist movement has been extremely aggressive in using the courts to suppress the rights of creationists to express their viewpoint, even to the point of forbidding school teachers and university lecturers from offering a creationist model as an alternate theory. Dr. John G. West, with the Discovery Institute's Centre for Science and Culture, stated:

> *"Darwinist efforts to use the courts, the media and academic tenure committees to suppress dissent and stifle discussion are in fact fueling even more dissent and inspiring more scientists to ask to be added to the list [of dissenters]."*[12]

While evolutionist documentaries are given prime time coverage in media outlets, material produced by creationists is often subject to censorship and shockingly disparaging campaigns. For example, the 2008 documentary, *"The Privileged Planet"*, was a beautifully produced creationist film documenting the extraordinary evidence for intelligent design, based upon the highly improbable "coincidence" of life-permitting factors that enables our planet to be life-permitting.

Creation Ministries International describes the horrible campaign that was waged by atheists against the film:

> *"The Smithsonian Institute in the United States had initially agreed to provide a private screening, but then attempted to cancel it following widespread protests. Atheist groups organised campaigns, encouraging people to send e-mails, write letters and make phone calls opposing its showing. An e-mail was sent to the entire department of anthropology at George Washington University warning everyone not to watch it. Money was even offered to the Smithsonian if they agreed not to show it. Dr Gonzalez himself received much criticism, leading to his promising career at Iowa State University coming to an abrupt end and his being denied tenure—a situation which can have grave implications for a person's future in academia."*[13]

This represents a blatant shutting down of dissenting viewpoints by the atheistic evolutionist movement. The British Humanist Association, as another example, has successfully lobbied the British government to censor any mention of creationism or evidence of intelligent design in any classes in state schools, even in Religious Education classes.[14]

These kind of draconian measures, are accurately described by Dr. Henry Morris as *"the long war against God"*.[15] Dr. Morris is absolutely correct. There is a spiritual battle taking place in our world. Underlying this ugly, hostile persecution and censorship of Christians by evolutionists, is a dark spiritual opposition to the rule of God and his claim upon our lives. It is an intransigent unwillingness to recognise his rulership over us, and to proclaim ourselves as kings, instead.

Jeremy Rifken, the American evolutionist and atheist, in his book, "*Algemy*", wrote:

> *"We no longer believe ourselves to be guests in someone else's home and therefore obliged to make our behaviour conform to a set of pre-existing cosmic rules. It is our creation now; we make the rules, we establish the parameters, we create our own world; and because we do, we no longer have to justify our behaviour. We are the architects of the universe, we are responsible to nothing outside ourselves, for we are the kingdom, the power and the glory forever and ever!"*[16]

This is why those who have rejected God will oppose his representatives with equal vigour. Jesus warned that this would be the case:

> *"If they persecuted me, they will persecute you also."* (John 15:20)

**EVOLUTION AS RELIGION**

Make no mistake; belief in evolution is a religion. Evolution is not an unbiased scientific theory, but a whole belief system based on the premise that there is no God and that we are masters of our own destinies. For atheists, evolution is the new god that has replaced the God of the Bible. Evolutionist philosopher, Teilhard de Chardin, wrote:

> "[Evolution] is a general postulate to which all theories, all hypotheses, all systems must henceforward <u>bow</u> and which they must satisfy in order to be thinkable and true. Evolution is a light which illuminates all facts, a trajectory which all lines of thought must follow."[17]

Underlying this attitude is what can only be described as a philosophy of pantheism; belief in, and worship of, a semi-conscious divine power that is at work within the "evolutionary universe". Jeremy Rifkin, once again, states:

> "Evolution is no longer viewed as a mindless affair, quite the opposite. It is <u>mind</u> enlarging its domain up the chain of species[18] ... In this way one eventually ends up with the idea of the universe as <u>a mind that oversees, orchestrates, and gives order and structure to all things.</u>[19]" [Underlining mine]

Here is the religious philosophy of evolution in all its glory. It is a belief in the "mind" of the universe to create life in its own image. It is a belief in a Star Wars-like power that pervades the universe and implements its sub-conscious will upon the physical world. This quasi-religious underpinning of evolutionary theory, at least by some, reveals that when the true God is removed from consideration, mankind has an inherent need to replace him with something less offensive to our own self-proclaimed sovereignty. An impersonal pantheistic power is seen by some as far more tolerable than a God who demands our repentance and our worship.

## A BATTLE OF TWO RELIGIONS

Thus, the evolution / creation debate is not, at its heart, a clash of two scientific theories. Nor is it, as evolutionists would have us believe, a dispute between science and religion. Fundamentally, it is a clash between two diametrically opposed religious worldviews. On the one hand, there is Evolution, with mankind as the supreme ruler of creation, benignly assisted by an insipid, nebulous, cosmic power. On the other hand, there is Christianity, with God as the sovereign King over creation, and mankind subservient and accountable to him.

Richard Bossard, an outspoken atheist and evolutionist, stated:

> *"The day will come when the evidence for evolutionary theory becomes so massively persuasive that even the most fundamental Christian warriors will have to lay down their arms and surrender unconditionally. I believe that day will be the end of Christianity!"*[20]

Contrary to this view, however, the tide of the battle has turned in the opposite direction. Today, more than ever before, there is significant and growing evidence contradicting evolutionary theory and supporting the biblical account of creation.

---

**ENDNOTES - Chapter 18**

---

[1] Dawkins R., *The Blind Watchmaker*, Penguin, London, p. 6, 1991

[2] D.M. Walton, London Times, August 3, 1929, cited by Bolton Davidheiser, "Evolution and Christian faith", Grand rapids, Baker Book House, 1969, p.155.

[3] Nagel, Thomas, The Last Word, pp. 130–131, Oxford University Press, 1997. Dr Nagel (1937– ) is Professor of Philosophy and Law at New York University.

[4] Edward Feser, "The Last Superstition", St. Augustine's Press, 2008, p.31

[5] Wald, George, "Innovation and Biology," Scientific American, Vol. 199, Sept. 1958, p. 100

[6] Originally posted by Richard Dawkins on his website. It has subsequently been removed, but unfortunately for Dawkins, the damage was already done. The quote was picked up by various critics, and is published in a number of online articles. For example:

"Is Richard Dawkins Really That Naive?", by Massima Pigliucci, on https://www.psychologytoday.com/au/blog/rationally-speaking/200904/is-richard-dawkins-really-naive.

[7] David Klinghoffer, "Meet Gunter Bechley", https://evolutionnews.org/2017/12/meet-gunter-bechly/

[8] https://www.haaretz.com/science-and-health/scientist-comes-out-against-evolution-loses-wikipedia-page-1.5466166

[9] https://www.haaretz.com/science-and-health/scientist-comes-out-against-evolution-loses-wikipedia-page-1.5466166

[10] John F. Ashton, "In Six Days: Why Fifty Scientists Choose To Believe in Creation", Green forest, AR, Masterbooks, 2001, p.22.

[11] Directed by Nathan Frankowski, "Expelled: No Intelligence Allowed", 2008.

[12] John G. West, quoted in "Over 500 Scientists Proclaim Their Doubts About darwin's Theory of Evolution", https://evolutionnews.org/2006/02/over_500_scientists_proclaim_t/

[13] "No Coincidence", an online article on creation.com/no-coincidence

[14] Statham, D.R., Evidence for Creation now banned from UK religious education classes; creation.com/creation-religious-education

[15] Henry Morris, cited in "No Coincidence", an online article on creation.com/no-coincidence

[16] Jeremy Rifkin, *Algeny* (New York, Viking Press, 1983), p. 244

[17] Cited in "Nothing in Biology Makes Sense Except in the Light of Evolution: Theodosius Dobzhansky, 1900-1975," by Francisco Ayala, in *Journal of Heredity* (vol. 68, no. 3, 1977), p. 3.

[18] Jeremy Rifkin, *Algeny* (New York, Viking Press, 1983), p. 188.

[19] Jeremy Rifkin, *Algeny* (New York, Viking Press, 1983), p. 195.

# The Atheistic Agenda

[20] Quoted by Ken Ham, on the "Caister Seminar, UK 1996" audio tapes by Ken Ham, distributed by Creation Science Foundation.

# Evolution and the Bible

# Chapter 19

# Evolution and the Bible

Some people believe that evolution and the Bible can be blended together; that it is possible to believe both at the same time. A significant number of Christians believe that God used evolution to create biological life. This view is called "theistic evolution". A survey of American churches in 2007 revealed that 51% of Protestants and 58% of Catholics believe that God used evolution to create life.[1] A more recent survey by the Pew Research Centre in 2018 revealed that belief in evolution among American protestants has increased significantly in the last decade:[2]

- 58% of white Evangelicals
- 66% of black Protestants
- 56% of Catholics

# Evolution and the Bible

The attempt to combine evolution and the Bible is based upon the naive assumption that evolution is a proven fact and that it is our interpretation of the Bible that needs to be modified in order to accommodate the more authoritative truth of science. It is remarkable how quickly and easily some branches of the Christian church have bowed to the pseudo-science of evolution and rushed to reinterpret the Bible. Consider the following:

- In 1996, Anglican Archbishop of Brisbane, Peter Hollingworth, wrote the foreword to evolutionist Ian Plimer's book, "*Telling Lies For God*". The book was a scathing attack upon the doctrine of creation and upon scientists with a Christian faith who seek to defend it. In his foreword, Rev. Hollingworth stated: *"The theory of evolution is a more sound explanation of the origin of our world than the creation story which is based upon Biblical literalism."*[3]

- In 1982, the Episcopal Church passed a resolution to *"affirm its belief in the glorious ability of God to create in any manner, and in this affirmation reject the rigid dogmatism of the 'Creationist' movement."*[4]

- The Evangelical Lutheran Church in America recently issued a statement that *"God created the universe and all that is therein, only not necessarily in six 24-hour days, and that God actually may have used evolution in the process of creation."*[5]

- In 2008, the United Methodist Church's General Conference issued a statement that *"science's descriptions of cosmological, geological, and biological evolution are not in conflict with [the church's] theology."*[6]

- The United Church of Christ, in America, has recently issued a statement indicating that they were *"embracing evolution as a means to see our faith in a new way."*[7]

I could go on, but I think you get the picture! In recent decades, denominational leaders have practically been knocking each other over in their rush to embrace evolution and reinterpret the Bible!

Many Christians believe that combining evolution and the Bible is a simple matter of reinterpreting relatively few verses, but this is not the case at all. This chapter will highlight the huge disparity between the Bible and evolution, and will demonstrate that it is not possible to believe both simultaneously.

# REASONS WHY EVOLUTION AND THE BIBLE ARE INCOMPATIBLE

### 1. NO DEATH BEFORE ADAM AND EVE'S FALL

Evolution teaches that animals had been living and dying for millions of years before humans evolved on our planet. Death was supposedly a part of the world for millions of years before humans, as animals killed and devoured each other. The Bible, however, teaches that there was no death in the world until Adam sinned:

> "Sin entered the world through one man [Adam], and death through sin, and in this way, death came to all people, because all sinned." (Rom 5:12)

> "For since death came through a man, the resurrection of the dead comes also through a man. For as in Adam all die, so in Christ all will be made alive." (1 Cor 15:21-22)

Some people argue that these verses only refer to human death, not animal death. In other words, they postulate that animals may have been killing and devouring each other, and dying of "natural" causes, long before mankind sinned. But the use of the word, "world", in Romans 5:12 (above) does not allow that interpretation. The Bible is very clear; before

Adam sinned, there was no death in *"**the world**"*. This is why God was able to say of his perfect world when he had finished creating it, *"It was very good" (Gen 1:31)*. This is not just a minor detail; it is a huge discrepancy, because evolution teaches that the earth was a place of death and violence and bloodshed as species fought for dominion for millions of years.

## 2. ADAM WAS MADE FROM DUST

According to Genesis, Adam was made from the *"dust of the ground"* (Gen 2:7), and Eve from Adam's rib (Gen 2:21-23). Evolution, on the other hand, proposes that humans evolved from other animals. Some evolutionised Christians argue that "dust" represents the chemicals that God used to start the evolutionary process. Thus "dust-to-Adam" represents "chemicals-to-humans". Yet, if this is true, what does "rib-to-Eve" represent? To be consistent, one needs an adequate explanation - and there is none. Eve did not come from a dust cocktail of chemicals, but from an already fully functional created man.

Other people say that the "dust" in Genesis 2:7 represents the ape-like creatures from which God formed man. But, again, one must be consistent. Genesis 3:19 tells us that *"dust you are, and to dust you will return."* According to the Bible, when we die, we return to the dust from which we were made. Clearly, we do not return to being ape-like creatures when we die! We decompose and return to the basic elements in the soil of the earth - the same soil that we were made from. This is an important point. Mankind was created by God's hands from the dust of the earth in a single day, not from apes over millions of years.

## 3. CREATION IS FINISHED

Evolutionists believe that evolution is a continual process; that it is still occurring today. If it is not still happening, they argue, then there is no basis for extrapolating into the past to say that it happened back then. Evolution supposes that the evolutionary process is ongoing, and many evolutionists maintain that in the distant future a higher lifeform will evolve, for whom we are an intermediate stepping stone. In other words, creation is never finished; it is continually evolving. Yet the Bible is very clear that the work of creation is finished. "*Thus the heavens and the earth were completed in all their vast array. By the seventh day God had finished the work he had been doing; so on the seventh day he rested from all his work. And God blessed the seventh day and made it holy, because on it he rested from all the work of creation that he had done.*" (Genesis 2:1-3).

## 4. JESUS BELIEVED GENESIS

Many evolutionised Christians maintain that Genesis is only symbolic - a kind of analogy. According to this view, the events described didn't actually occur, but are simply myths with a meaning. Yet Jesus expressed his belief in the literal accuracy of Genesis, and portrayed the Genesis record as foundational for the Christian faith. In John 5:6-7, he said "*If you believed Moses you would believe me..... But since you do not believe what he wrote, how are you going to believe what I say?*" This reference to believing "Moses" is a reference to the writings of Moses, the first five books of the Bible; Genesis, Exodus, Leviticus, Number and Deuteronomy.

In his teaching on marriage, in Matthew 19:4-6, Jesus quotes the creation of Adam and Eve by God, in Genesis chapter 2, as foundational for our concept of the sanctity and holiness of marriage in God's eyes. He states:

> "*Haven't you read,*" *he replied,* "*that at the beginning the Creator 'made them male and female, and said, 'For this reason a man will leave his father and mother and be united to his wife, and the two will become one flesh'? So they are no longer two, but one flesh. Therefore what God has joined together, let no one separate.*"

# Evolution and the Bible

This is a direct quote from Genesis 1:27 and 2:24, which describe, in great detail, God's creation of the first two human beings, Adam and Eve. There is not even the slightest hint in Jesus' words that he considers the creation of Adam and Eve to be anything other than literal human history. He specifically appeals to the creation of these first two people as the universal, binding basis for monogamous heterosexual marriage. If Adam and Eve are a myth, it makes a mockery of the whole Christian doctrine of monogamous marriage.

It is also important to point out that Jesus stated that God created mankind "at the beginning". However, on the evolutionary story and time scale, and using the common allegory of earth's supposed 4.5 billion year history, being approximately represented by a single 24 hour day; the first single cellular life appears at about 4:00 am, seaweed at am, marine animals not until 10:00 pm, then dinosaurs at 11:00 pm, mammals proliferate around 11.30 pm, and people don't appear until about a second before midnight. And that doesn't even allow for another supposed nine and a half billion years or so, before the earth formed. In the evolutionary story, people are clearly not made at the beginning. This is another huge contradiction between the Bible and evolutionary theory. Christians must ask themselves, *"Whom will I believe: Jesus or fallible human scientists?"*

Another reference to Genesis by Jesus is found in Luke 11:51 and Matthew 23:35, where Jesus states:

> *"... from the blood of Abel to the blood of Zechariah, who was killed between the altar and the sanctuary. Yes, I tell you, this generation will be held responsible for it all."*

Here, Jesus is referring to Abel, Adam's son, as a real person, and telling the teachers of the law, the Pharisees and *"this generation"* that they will be held accountable for Abel's murder. It would seem odd to share in the guilt for the murder of a non-existent person!

Another reference to Genesis by Jesus is found in Matthew 24:36-39 (and Luke 17:26-27), where he refers to the worldwide flood of Noah, recorded in Genesis 6-9. He states:

> *"But about that day or hour no one knows, not even the angels in heaven, nor the Son, but only the Father. As it was in the days of Noah, so it will be at the coming of the Son of Man. For in the days before the flood, people were eating and drinking, marrying and giving in marriage, up to the day Noah entered the ark; and they knew nothing about what would happen until the flood came and took them all away. That is how it will be at the coming of the Son of Man."*

Once again, this is not a mythological story that Jesus is referring to. His detailed description of the sinful revelry of humanity *"up to the day Noah entered the ark"*, followed by the dreadful description that *"the flood came and took them all away"* leaves us in no doubt that Jesus regards the global flood as a literal historical event. Indeed, his vivid description of these events is that of an eye-witness, which of course he was, as the eternal Son of God. Furthermore, Jesus' statement that *"this is how it will be at the coming of the Son of Man"* (a reference to his impending second coming) is significant. If we relegate the flood to the realm of myth, it necessarily drags the concept of the second coming into that same realm. Yet the New Testament is unequivocal that Christ WILL one day return in the culminating act of history, when *"every eye will see him and every knee will bow"* (Rev 1:7). The literal return of Christ will involve the judgment of all humanity, just as took place in the literal global flood of Noah.

Some people say that these references by Jesus to the events of Genesis were simply a concession that Jesus made to the popular myths of his day; that he knew they weren't true, but referred to them simply because they illustrated his teachings. But the Bible teaches that Jesus is the

TRUTH (John 14:6). To say that Jesus would knowingly teach myth as fact, is to call him a liar. Jesus clearly believed Genesis to be actual history. He believed Adam and Eve to be actual history. He believed Noah's flood to be literal history. Do we dare believe any less?

## 5. ALL PEOPLE ARE DESCENDED FROM ADAM

According to the Bible, all human beings are descended from Adam and Eve. Acts 17:26 says, *"From one man, God made all the nations of the earth."* Genesis 3:20 says, *"Adam named his wife Eve, because she would become the mother of all people"*. This contradicts evolutionary theory which proposes that there was not just one first man and woman, but many, as a growing number of ape-like creatures made the transition from animal to human. Some evolutionised Christians try to reconcile these contradictory teachings by maintaining a belief in a literal Adam and Eve, claiming that they were just two individuals of the many who made the transition from the animal kingdom. If this is so, then some of us would be descended from Adam, and some would be descended from the other Adam-like first humans. Adam would, therefore, be the representative of only some of the human race, and not all of us, as the Bible clearly teaches (Rom 5:12).

In Chapter 10, "Meet Your Ancestors", we examined the genetic evidence indicating the existence of Mitochondrial Eve and Y-Chromosomal Adam - genetic progenitors from whom every person living today is descended. This scientific evidence strongly supports the biblical account of the creation of mankind from just two original humans; Adam and Eve.

## 6. JESUS IS DESCENDED FROM ADAM

Many theistic evolutionists believe that Adam was not an actual person; that Adam and Eve are just mythological people representing the first generation of humans who evolved from pre-human creatures. Yet Jesus and the gospel writers traced Jesus' ancestry back to Adam (Matt 1:1-16;

Lk 3:21-38). For example, Luke presents a very clear genealogy of Jesus, all the way back to Adam:

> *"Now Jesus himself was about thirty years old when he began his ministry. He was the son, so it was thought, of Joseph, the son of Heli, the son of Matthat, the son of Levi, .... the son of Joshua, the son of Eliezer, .... the son of Levi, the son of Simeon, the son of Judah, the son of Joseph, .... the son of Nathan, the son of David, the son of Jesse, the son of Obed, the son of Boaz, .... the son of Judah, the son of Jacob, the son of Isaac, the son of Abraham, the son of Terah, .... the son of Shelah, the son of Cainan, the son of Arphaxad, the son of Shem, the son of Noah, the son of Lamech, the son of Methuselah, the son of Enoch, .... the son of Kenan, the son of Enosh, the son of Seth, <u>the son of Adam, the son of God</u>." (from Luke 3:23-38)*

If Adam was a myth, is Jesus also a myth? We have already examined the strong historical evidence for the existence of Jesus. So, if we accept Jesus as historical, where in these genealogies does myth stop and fact start? Once again, we make Jesus and the gospel writers out to be liars if we claim that Adam did not exist.

## 7. DAYS CANNOT BE MILLIONS OF YEARS

Evolutionised Christians claim that the Genesis "days" of creation represent millions or billions of years of distinct evolutionary periods. There are many problems with this hypothesis.

**Firstly**, the Genesis account clearly defines "*day*" for us. The Hebrew word for day, "*yom*", can certainly mean an indefinite period of time. For instance, the Bible sometimes refers to the "*day of the judges*", meaning an indefinite period of time longer than a normal day. But Genesis chapter 1 clearly rules out this interpretation by defining each day as one morning and one evening:

*"There was evening, and there was morning - the first day" (Gen 1:5).*
*"There was evening, and there was morning - the second day" (Gen 1:8).*
And so on.

The specific numbering of the days of creation in Genesis chapter 1 is also significant: *"the first day ... the second day ..."*, and so on. The creation account does not say, *"There was evening and there was morning, the first undefined period of time.... There was evening and there was morning, the second undefined period of time."* In fact, the scriptures could not be clearer at this point. It's as if God has gone out of his way to ensure that we get the message, loud and clear, that these days of creation were, indeed, 24-hour days. It defies all rules of grammar and common sense to interpret these days as millions or billions of years.

**Secondly**, Exodus 20:11 tells us that God created everything in six days and rested on the seventh as a pattern for man to follow. He now commands us to do the same (Exod 20:9-10). It would be ludicrous to suggest that God created in six billion years and rested for one billion years, and tells us to do the same! It makes even less sense to suggest that he created for six indefinite periods of time and rested for one indefinite period of time, and commands us to follow his pattern. Genesis 1 and Exodus 20 are CLEARLY referring to normal days.

**Thirdly**, Adam was created on day six and lived through days six and seven (Gen 1:27-2:3). To suggest that each day was millions of years makes Adam's existence through the last two days of that first week an absurdity (particularly when Gen 5:5 states that Adam lived for 930 years).

## 8. GENESIS AND EVOLUTION HAVE A DIFFERENT SEQUENCE

The following table shows how completely irreconcilable are evolution and the Genesis record:

| EVOLUTION | GENESIS |
|---|---|
| Stars formed before the earth | The earth was formed first (Gen 1:6) |
| Light resulted from the formation of stars | Light was created 3 days before the sun, moon and stars (Gen 1:3,14) |
| Aquatic life, protozoa and metazoa were formed before complex plant life | All plant life was formed before any other life forms (Gen 1:9-13) |
| Amphibia and reptiles preceded birds by millions of years | Birds were created before any land animals (Gen 1:20-23) |
| The universe was formed over billions of years | The universe was created in 6 days |

## 9. ADAM WAS NOT PRIMITIVE

Evolutionists teach that the first humans were extremely primitive; speaking in grunts and having no advanced technological skills. Yet the Bible describes Adam and his offspring as having a highly developed language (Gen 2:20), using advanced farming methods (Gen 4), playing musical instruments (harp and flute - Gen 4:21), and forging tools out of bronze and iron (Gen 4:22). According to evolution, these developments took hundreds of thousands, or even millions of years, yet the Bible shows them as being in evidence with the first human beings.

Furthermore, it is clear from the Genesis account that Adam had extremely high functioning cognitive abilities. The fact that he named all the animals as God presented them to him, **and accurately retained the memory of all those names**, is an indication of great intelligence. This accords with the concept that in their pre-fall state, Adam and Eve were perfect; not yet subject to the deteriorating consequences of the fall.

## 10. GENESIS IS THE FOUNDATION FOR MANY NEW TESTAMENT DOCTRINES

The writers of the New Testament based key doctrines on the foundation of a literal interpretation of Genesis.

In Hebrews 11, the whole concept of faith is taught by referring to great men and women of faith from the past:

> *"Now faith is confidence in what we hope for and assurance about what we do not see.* [2] *This is what the ancients were commended for.*
>
> *By faith we understand that the universe was formed at God's command, so that what is seen was not made out of what was visible.*
>
> *By faith Abel brought God a better offering than Cain did. By faith he was commended as righteous, when God spoke well of his offerings. And by faith Abel still speaks, even though he is dead.*
>
> *By faith Enoch was taken from this life, so that he did not experience death: "He could not be found, because God had taken him away." For before he was taken, he was commended as one who pleased God. And without faith it is impossible to please God, because anyone who comes to him must believe that he exists and that he rewards those who earnestly seek him.*
>
> *By faith Noah, when warned about things not yet seen, in holy fear built an ark to save his family. By his faith he condemned the world and became heir of the righteousness that is in keeping with faith.*
>
> *By faith Abraham, ... Isaac and Jacob, ... by faith even Sarah, who was past childbearing age, was enabled to bear children ...*

> *All these people were still living by faith when they died. They did not receive the things promised; they only saw them and welcomed them from a distance, admitting that they were foreigners and strangers on earth." (From Hebrews 11:1-13)*

Please note that this list of faithful ancestors includes Abel, Adam's son, and many others from the succeeding generations. There is no hint that any of these people were anything but literal human beings who served God faithfully, and whom we should seek to emulate. If they were fictitious, they could hardly be an inspiration for us to follow!

In his first epistle, the Apostle Peter describes an intriguing aspect of the post-resurrection ministry of Jesus:

> *"After being made alive, he went and made proclamation to the imprisoned spirits— <u>to those who were disobedient long ago when God waited patiently in the days of Noah while the ark was being built. In it only a few people, eight in all, were saved through water,</u> and this water symbolizes baptism that now saves you also—not the removal of dirt from the body but the pledge of a clear conscience toward God. It saves you by the resurrection of Jesus Christ, ..." (1 Peter 3:19-21)*

If the flood of Noah was only a myth, who did Jesus preach to in this encounter? This verse makes no sense at all unless there was a literal flood, a literal Noah and a literal generation of disobedient people who were judged by God *"while the ark was being built"(v.20).*

In Romans 5, Paul describes Jesus as the "*second Adam*", and shows that it was Adam's sin which led to all mankind inheriting a sinful nature and, therefore, needing to be redeemed. Jesus, the second Adam, lived the perfect life of obedience that Adam failed to live. Jesus therefore earned the right to represent us on the cross and purchase our forgiveness.

Similarly, Paul writes:

> *"For since death came through a man, the resurrection of the dead comes also through a man. For as in Adam all die, so in Christ all will be made alive. But each in turn: Christ, the firstfruits; then, when he comes, those who belong to him."* (1 Corinthians 15:21-23)

Note: Adam was a *"man"* not a myth. It is abundantly clear that the New Testament writers regarded Adam as a literal person and the precursor to the saving work of Christ.

If there was no literal Adam, it makes a mockery of this whole doctrine. Without a literal Adam and a literal rebellion in the garden, there is no original sin and no need for redemption. This makes the life and death of Jesus a pointless sacrifice.

The outspoken atheist, Dr. Richard Dawkins, pointed out the absolute essential nature of a literalist view of Adam and Eve to the Christian doctrine of Christ's atonement, when he wrote:

> *"To cap it all, Adam, the supposed perpetrator of the original sin, never existed in the first place: an awkward fact – excusably unknown to Paul but presumably known to an omniscient God (and Jesus, if you believe he was God?) – which fundamentally undermines the premise of the whole tortuously nasty theory. Oh, but of course, the story of Adam and Eve was only ever symbolic, wasn't it? Symbolic? So, in order to impress himself, Jesus had himself tortured and executed, in vicarious punishment for a symbolic sin committed by a non-existent individual? As I said, barking mad, as well as viciously unpleasant."*[8]

Here, Dawkins is pointing out the absurdity of Christians believing in the atonement of Christ while simultaneously regarding Adam as mythological. He understands the essential, irrevocable link between a literal Adam and the atonement of Christ better than many Christians do!

Many other key doctrines in the New Testament are based upon a literal, historical interpretation of the events of Genesis (e.g.: 1 Cor 15; Matt 12; Matt 16; 1 Pet 3; Luke 11:29f) In fact, Adam or Eve are mentioned 13 times in the New Testament as real historical people. Dr. Ken Ham points out:

> *"There are at least 165 passages in Genesis that are either directly quoted or clearly referred to in the New Testament. Included in these are more than 100 quotations or direct references to Genesis chapters 1 to 11."*[9]

**IRRECONCILABLE DIFFERENCES**

The ten issues listed above indicate how completely irreconcilable evolutionary theory is with the Biblical account of creation. It is simply not possible to believe both simultaneously without having to ignore or reinterpret large sections of the Bible.

Thomas Huxley, the famous evolutionist and humanist, and a contemporary of Charles Darwin, once gave a lecture on the incompatibility of evolution and Christianity. He stated:

> *"If Adam is a myth, what value is Paul's teaching in Romans 5? And what about the authority of the writers of the New Testament, and of Jesus himself, who, on this theory, have not merely accepted flimsy fiction as solid facts, but have built the very foundations of Christian doctrine upon legendary quicksand!"*[10]

Thomas Huxley was referring to the tendency of some Christians to reinterpret the Bible to fit in with evolution. He saw that many Christians were treating the Genesis account as non-literal in order to accommodate the Bible to the evolutionary narrative. Huxley's response was to effectively say to Christians, *"You can't do that! If you treat your creation account as mythological, you undermine the rest of the Bible!"*

Huxley was an atheist, yet he saw clearly what many Christians do not; that the Bible and evolution are completely incompatible.

Richard Bossard, an outspoken evolutionist and atheist, made a similar comment about the irreconcilable differences between evolution and biblical Christianity:

> *"Christianity will fight evolution to the bitter end because evolution destroys utterly and finally the very reason why Jesus' earthly life was supposedly necessary. Destroy Adam and Eve and the original sin, and in the rubble, you will find the sorry remains of the Son of God. Evolution takes away the meaning of his death. If Jesus was not the second Adam who died for our sins (and this is what evolution means), Christianity means nothing!"*[11]

## CONCLUSION

There is no way of reconciling the vast differences between the teachings of Genesis and evolution. People who try to do so are ignoring the many mutually contradictory teachings and, in the end, distorting and misinterpreting the Bible to make it "fit" with evolutionary theory. If you choose to believe in evolution, you must then disregard large portions of the Bible. Not only must Genesis be dismissed as myth, but also many key biblical doctrines lose their whole foundation.

Many secular historians and scholars seem to understand the irreconcilable differences between the Bible and evolutionary theory better than many Christians. Dr. James Barr, a professor at Oxford University, England, wrote, in a letter to David C. Watson, on 23rd April 1984:

> *"So far as I know, there is no professor of Hebrew or Old Testament at any world-class university who does not believe that the writer(s) of Genesis 1–11 intended to convey to their readers the ideas that: (A) Creation took place in a series of six*

> *days which were the same as the days of 24 hours we now experience; (B) The figures contained in the Genesis genealogies provided by simple addition a chronology from the beginning of the world up to later stages in the biblical story; and (C) Noah's flood was understood to be world-wide and extinguish all human and animal life except for those in the ark.*"[12]

It should be noted that Dr. Barr did not, himself, believe Genesis 1-11 to be literally true, but he was adamant that the text itself does not allow for any other interpretation. As he further explained in the same letter to David Watson, it was only the perceived need to harmonise the Genesis creation account with the new evolutionary narrative which subsequently led Christians to reinterpret these early chapters of the Bible; but this reinterpretation has no basis in the text itself.

We must choose between the Bible and evolutionary theory, because they cannot both be true. There is a big issue at stake here: Will we trust the imperfect and ever-changing theories of fallible scientists, or will we trust the unchanging Word of God, which is "perfect"?

> *"The law of the Lord is perfect"* (Psa 19:7)

Whom will you believe? Fallible humans who were not there at the beginning? Or the God of the universe, who *was* there, and who has revealed the truth to us in his Word?

> *"Where were you when I laid the foundation of the earth? Tell Me, if you have understanding."* (Job 38:4)


**A FINAL QUOTE:**

Dr. David Berlinski earned his Ph.D. in philosophy from Princeton University and was later a post-doctoral Professor of Mathematics and Molecular Biology at Columbia University. He wrote:

*"Darwin's theory of evolution is the last of the great nineteenth-century mystery religions. As we speak, it is now following Freudians and Marxism into the nether regions, and I'm quite sure that Freud, Marx and Darwin are commiserating one with the other in the dark dungeon where discarded gods gather."*[13]

---

**ENDNOTES - Chapter 19**

---

[1] Survey conducted by "The Pew Forum", 2007, published 2008.

[2] Survey of American churches by Pew Research Centre, 2018, published in "Christianity Today", April 2019, "An Evolution Poll Changes Over Time.

[3] Peter Hollingworth, foreword to Ian Plimer, "Telling Lies For God", Random House, Australia, 1994,

[4] The Archives of the Episcopal Church, *The Acts of the Convention 1976-2006*, "Reject the Dogma of Creationism"

[5] http://www.pewforum.org/2009/02/04/religious-groups-views-on-evolution/

[6] United Methodist Church, General Conference 2008, Legislation Tracking, "Science and Technology, Petition 80050"

[7] John H. Thomas, United Church of Christ, "A New Voice Arising: A Pastoral Letter on Faith Engaging Science and Technology", Jan 2008.

[8] Richard Dawkins, "The God Delusion", Bantam Books, 2006, p.287

[9] https://answersingenesis.org/theistic-evolution/twenty-reasons-why-genesis-and-evolution-do-not-mix/

[10] Thomas Huxley, quoted in https://creation.com/christianity-stands-or-falls

[11] Richard Bossard, quoted by Ken Ham, From the "Caister Seminar, UK 1996" audio tapes by Ken Ham, distributed by Creation Science Foundation.

[12] Dr. James Barr, in a letter to David C. Watson, 23rd April 1984, quoted in https://creation.com/oxford-hebraist-james-barr-genesis-means-what-it-says

[13] https://www.azquotes.com/quote/589850

# OTHER TITLES BY KEVIN SIMINGTON

"***Finding God When He Seems To Be Hiding***" provides intelligent answers to the common questions and objections that are often roadblocks in people's journey towards faith. If God wants people to believe in him, why doesn't he make himself more obvious? What evidence is there for God's existence? If God is so good, why is there so much suffering in the world? God-ordained Christian killing in the Bible makes Christians no better than terrorists! What sort of narcissistic God eternally tortures people for not loving him? Is the Bible reliable? Has it changed over time? Is the life of Jesus a myth? What evidence is there for his resurrection? What about evolution? Hasn't science and evolution disproved the existence of God? How can God permit abuse and religious violence?

These and other questions can bring into doubt the very existence and character of God, and demand answers that move beyond the standard, glib responses that are often proposed. This book addresses these challenging issues with remarkable clarity and insight. It proposes meaningful answers that will enable earnest seekers and puzzled believers to develop an unshakable confidence in the unimpeachable character of God.

"***Finding God When He Seems To Be Hiding***" is available in print or as an eBook from SmartFaith.net and from all major retailers.

**This book will change the way you read the Bible!**

"*Making Sense of the Bible*" is a comprehensive guide to understanding and interpreting the Bible. It begins by examining the remarkable journey of the Bible, from original text to modern translation, and will assist you to develop a more mature, complex understanding of the nature of its divine inspiration. It then examines the many complex cultural and contextual issues that are essential in order to accurately apply the Bible's message. These include the difference between the two covenants, the nature of progressive revelation, the pre-Christian context of the Old Testament, and the necessity to read the whole Bible "Christologically" - through the lens of Christ's person and work.

What sets "*Making Sense Of The Bible*" apart from similar books is its intensely practical nature. Commonly misinterpreted doctrines are explored in detail, and important principles of interpretation are applied. A large range of key biblical doctrines are examined in detail.

**This book is a must for ordinary Bible readers and serious students alike!**

"*Making sense of the Bible*" is available in print or as an eBook from SmartFaith.net and from all major retailers.

# CONNECT WITH KEVIN SIMINGTON

Visit Kevin Simington's website:

https://smartfaith.net

 Kevin Simington
SmartFaith.net

SmartFaith.net contains a large repository of helpful resources in the areas of apologetics, theology and philosophy.

Subscribe to Kevin's weekly blog at https://smartfaith.net/blog/

Visit Kevin's facebook page, "Reflections on Faith and Life" at https://www.facebook.com/ReflectionsKev/

## AUTHOR BIO

Kevin Simington (B.Th. Dip. Min.) is a theologian and apologist who is passionate about helping Christians to grow deeper in their faith. He spent 31 years in Christian ministry, as a church pastor and a Christian educator. He is now a full time author and speaker. His website, SmartFaith.net and Facebook page, "Reflections on Faith and Life", provide valuable resources for defending the Christian faith and equipping Christians. Kevin's weekly blog, available through his website and Facebook page, provides incisive commentary on social issues, theology, apologetics and ethics, and is read by thousands of people worldwide.

www.ingramcontent.com/pod-product-compliance
Lightning Source LLC
Chambersburg PA
CBHW071852290426
44110CB00013B/1117